Landfill Waste Pollution and Control

ABOUT OUR AUTHOR

KENNETH WESTLAKE is a member of the Institute of Waste Management, where his work on landfill control and techniques is respected world-wide. A graduate with a B.Sc.(Hons) in Applied Biology followed by M.Phil., both from Nottingham Trent University (formerly Trent Polytechnic). He read for his doctorate as Lecturer in Biochemistry in south Africa at the University of Natal and subsequently he became, in 1983, Agricultural Research Officer to the South African Animal and Dairy Science Research Institute.

His next appointment was as Senior Scientific Officer to the Institute of Food Research, Norwich, UK and Portland, Oregon, USA, with dual responsibility for an R&D programme investigating degradation processes in landfill. The objective of this assignment was to enhance landfill gas and energy production.

His next move was to become, within the Environmental Safety Centre at Harwell Laboratory from 1990-92, their Landfill Technology and Contaminated Land Consultant, and then Deputy Section Leader. Since 1992 he has been Lecturer in Hazardous Waste Management in the Centre for Hazard and Risk Management of Loughborough University of Technology. His responsibilities now include the teaching and administration of a postgraduate Hazardous Waste Management programme, and various industrial courses on Waste and Contaminated land-related subjects. He is a member of the Institute of Water Management.

LANDFILL WASTE POLLUTION AND CONTROL

KENNETH WESTLAKE, B.Sc.(Hons), M.Phil, Ph.D. M.Inst.W.M.
Waste and Environmental Group
Loughborough University of Technology
Loughborough

Albion Publishing
Chichester

First published in 1995 by
ALBION PUBLISHING LIMITED
International Publishers, Coll House, Westergate, Chichester, West Sussex,
PO20 6QL England

British Library Cataloguing in Publication Data
A catalogue record of this book is available from the British Library

ISBN 1-898563-08-X

Printed in Great Britain by Hartnolls, Bodmin, Cornwall

I hadn't intended to dedicate this book to anyone until my wife pointed out the error of my ways

TO BRIDGET

With Love

AUTHOR'S PREFACE

This book is intended as an overview of current policy, practice and management of landfills, that considers the relative merits of various options in each of the above, in the context of sustainable development. As such, I hope that those who are familiar with landfill will find food for thought, while those who have relatively little knowledge of landfill may gain a better understanding of some the **key principles and issues** in landfill today.

The sustainable landfill is as applicable to "developing" as to "developed" countries, although discussion of current policy, practice and management within this text relates to developed countries only - another book would be required to cover the same in developing countries. The book also focuses on the landfill of municipal solid waste rather than other particular waste categories, as it is often these wastes that provide the greatest problems; for these wastes, I attempt to compare policy in Europe and in the USA.

As first conceived, the book was to have a slightly different structure: changes have been made with the intention of providing a more fluent read. Thus the chapter on legislative control has been absorbed into the relevant sections elsewhere. Each chapter provides information on a particular element of landfill, as well as the necessary information to support the final chapter on the way forward for sustainable landfill development. It is argued that hazard reduction is the main area where landfill-associated risks can be reduced, and thus the gas and leachate-related hazards are explored in some detail, while the **principles** of landfill engineering are discussed, rather than the detailed practicalities. The book begins in Chapter 1, to provide a background to landfill, and to explore the concept of sustainable development, current landfill policy, and landfill as part of an integrated waste management strategy. In Chapter 2 the various principles of landfill management, (including landfill restoration and landfill mining) are introduced, while Chapter 3 explores the reactions that occur within the landfill; an understanding of these reactions is essential if we are to understand the associated hazards and risks, and hence manage landfill in a way that is sustainable. Chapters 4 and 5 look in more detail at the production, properties, migration and control of landfill gas and landfill leachate. An understanding of these parameters is critical to understanding how we may manage landfill in a more sustainable manner. Chapter 6 examines briefly other landfill associated hazards. Finally, Chapter 7 uses the information provided in the first 6 chapters to discuss the strengths and weaknesses in current waste and landfill management, and to explore the problems associated with the development of a sustainable landfill. Even if readers hold alternative opinions, then the views expressed may, at least, have encouraged new ideas and provided for healthy debate. We cannot continue to landfill waste in the way that we have been in the latter part of the 20th century without putting the environment at significant risk. The sooner we change to more sustainable practices the better!

The views expressed in this book have been formed, modified, and adopted as a result of discussion with many colleagues over the past five years. Their numbers are too numerous to mention, but to each I would like to express my thanks. I would also like to thank Judith Petts and David Campbell for helpful comments, my publisher Ellis Horwood, for patience and encouragement, and last but not least, my wife Bridget for early help in typing the text, but more especially for the support given during the (too) long time taken to write this book.

CONTENTS

Chapter 7   Sustainable Landfill - The Way Forward

LIST OF ABBREVIATIONS

BOD	Biochemical Oxygen Demand
BPEO	Best Practicable Environmental Option
CEC	Commission of the European Communities
CHP	Combined Heat and Power
COD	Chemical Oxygen Demand
COSHH	Control of Substances Hazardous to Health
CV	Calorific Value
DoE	(UK) Department of the Environment
DTi	Department of Trade and Industry (UK)
EC	European Commission
EPA	(UK) Environmental Protection Act (1990)
EU	European Union
ETSU	Energy Technology Support Unit (UK)
FML	Flexible Membrane Liner
HDPE	High Density Polyethylene
HRT	Hydraulic Retention Time
HSE	Health and Safety Executive (UK)
HSWA	Hazardous and Solid Waste Amendments (to RCRA)
ISWA	International Solid Waste Association
LCRS	Leachate Collection and Removal System
LEL	Lower Explosive Limit
LFMR	Landfill Mining and Reclamation
LLW	Low - level Radioactive Waste
MDPE	Medium Density Polyethylene
MSW	Municipal Solid Waste
MSWLF	Municipal Solid Waste Landfill
NFFO	Non - fossil Fuel Obligation
NIMBY	Not In My Back Yard (Syndrome)
NMOC	Non - methanogenic Organic Compound
NOx	Mixed Oxides of Nitrogen
NRA	National Rivers Authority (UK)
OEL	Occupational Exposure Limit
OES	Occupational Exposure Standard
PCB	Polychlorinated Biphenyl
RCEP	Royal Commission on Environmental Pollution
RCRA	Resource, Conservation, and Recovery Act
RDF	Refuse - derived Fuel
SOx	Mixed Oxides of Sulphur
STEL	Short - term Exposure Limit
TLV	Threshold Limit Value
TOC	Total Organic Carbon
TOE	Tonnes of Oil Equivalent
TON	Total Organic Nitrogen
UEL	Upper Explosive Limit
USEPA	United States Environmental Protection Agency
VFA	Volatile Fatty Acid
WMP	Waste Management Paper

GLOSSARY OF TERMS

Acetogenesis
The microbiological process that produces acetic acid - a key intermediate in the anaerobic degradation of waste within a landfill.

Anaerobic Digestion
An anaerobic biological process which, in the absence of oxygen, utilises biodegradable (organic) material for microbial growth. The end product is a relatively stabilised material, which has a lower organic content than the starting material. Anaerobic digestion is the main biological degradation process that occurs in landfill.

Attenuation
In this context, the reduction in toxicity of material as a result of natural processes.

Biodegradable
Material that may be degraded by biological activity.

BOD
A standard test for the presence of organic pollutants within liquids. It is, in effect, a measure of the content of biodegradable material.

COD
A measure of the amount of chemically oxidisable material in a liquid. It measures non-biodegradable, as well as biodegradable material.

Co-disposal
"the disposal, in landfills, of a restricted range of industrial wastes (including some special wastes) together with decomposing municipal waste or similar degradable wastes, in such a way that the industrial wastes gradually undergo a form of treatment" (DoE, 1994a).

Composting
An aerobic biological process in which biodegradable (organic) material serves as a source of "food" and energy for microbial growth. The end product is a relatively stabilised material, with a lower organic content, and which may be used beneficially - e.g. as a soil conditioner.

Field Capacity
The point at which waste becomes saturated. Once the field capacity is exceeded, free leachate will be generated.

Greenhouse Effect
The warming of the Earth's atmosphere caused by the release of certain gases which allow incident radiation to pass, but which "trap" reflected radiation, thus causing an increase in the temperature of Earth's atmosphere.

Groundwater
Water carried in underground rock bodies (aquifers).

Ion - exchange
One of the sorptive processes, occurring on surfaces in which weakly bonded ions are exchanged for ions that are more tightly bound.

Incineration

The thermal treatment of materials, normally under oxidising conditions, producing a low organic-content ash, and off-gases/particulates.

Integrated Waste Management

A system of waste management that utilises one, or more, of the available options (minimisation, recycling, re-use, incineration, digestion, composting, landfill etc.) for effective waste treatment and disposal.

Landfill

The controlled disposal of waste materials to land; differentiated from waste dumping in which no control is exercised.

Landfill - (Containment)

A landfill designed to contain liquids by the use of engineered liners. This type of landfill requires effective liquid management systems.

Landfill - (Dilute and Attenuate)

A landfill in which liquids are allowed to escape to the surrounding environment, and which recognises the attenuation potential of the surrounding environment.

Landfill - (Dry - tomb)

A landfill designed to prevent water infiltration, and within which waste remains relatively dry.

Landfill - (Failsafe)

A landfill designed and operated in such a way that when containment systems fail, or when there is loss of institutional control, the environmental risk is at an acceptable level.

Landfill - (Monodisposal) A landfill in which only wastes of similar characterisitics are emplaced. According to the nature of the wastes, attenuation may not occur and the mono-disposal landfill may act to store wastes in perpetuity.

Landfill - (Sustainable)

A landfill designed and operated in such a way that minimises both short-term and long-term environmental risks to an acceptable level.

Landfill Gas

A mixture of gases produced by the degradation of landfilled waste, and through volatilisation of emplaced waste and its degradation products. The major components are normally methane and carbon dioxide, although the gas mixture is normally comprised of more than 100 different gases.

Landfill Leachate

That liquid formed wihin a landfill site that is comprised of the liquids that enter the site (including rainwater), and the material that is leached from the waste and its degradation products, as the infiltrating liquids percolate downwards throught the waste.

Landfill Mining

The excavation of landfilled waste for product recovery.

Landraising

Waste disposal to land in which waste is emplaced above ground level.

Leachate Recycling

The collection of leachate from the base of a landfill and subsequent reintroduction in the upper layers of waste. The collected leachate may undergo treatment prior to reintroduction to the waste.

Methanogenesis

The microbiological process that produces methane gas. It is the final process in the degradation of organic waste within landfill.

Municipal Solid Waste

Waste collected and disposed by, or on behalf of a local authority. A large percentage of MSW will be household waste.

Refuse - derived Fuel

Processed MSW used as a fuel for boilers.

Sustainable Development

"development which meets the needs of the present without compromising the ability of future generations to meet their own needs" (From "the Brundtland Report").

Waste Management Hierarchy

A method of identifying appropriate waste management options, by listing the options in descending order of preference.

Waste Minimisation

The reduction in production of waste materials.

Waste Recycling

The re-use of waste materials after appropriate processing.

Waste Stabilisation

The process of degrading waste to produce a waste with a reduced pollution potential.

Water Balance

In this context, a calculation identifying liquid inputs to, and liquid losses from landfill. This facilitates landfill design and management to control liquid within the landfill environment.

CHAPTER 1

LANDFILL AND SUSTAINABLE DEVELOPMENT

1.1   Introduction

The disposal of waste to land has been an important method of waste disposal ever since the volume of waste arisings has been sufficiently great to warrant specific consideration. Since the late nineteenth century, the volume and the hazardous nature of waste generated has increased considerably, and has led to the need for disposal to land specifically allocated for the purposes of disposal - **landfill**. In the UK approximately 28 million tonnes of Municipal Solid Waste (MSW) are generated each year, 90% of which is disposed to landfill; 10 million tonnes is disposed to landfills with greater than 200,000 tonnes per annum capacity. In Italy, 74% of the 14 million tonnes of waste generated are disposed to landfill (Gendebien *et al*, 1992), and in the USA, a total of 195.7 million tons of MSW were generated in 1990, 66% of which was disposed to landfills (USEPA, 1992a). The design, operation, monitoring, and control of landfill, especially in 'developed' countries, is in a period of great change. Fundamental principles of landfill design and control are being challenged, and local, national, and international landfill policies and strategies are being rewritten in the light of increasing technical knowledge, and increasing public and pressure-group concern for the environmental harm that landfills may cause. However, it is clear that no matter what the landfill of the future will look like, it will continue to be needed as an important waste management option, for no matter how efficient waste minimisation and recycling schemes become, there will always be a proportion of the waste stream that requires disposal - love it or hate it, landfill disposal is here to stay!

Historically, the disposal of waste utilised poorly engineered 'tips' or 'dumps' in which no control over contamination of the environment was exercised. Until relatively recently, many landfills operated in this way, and it was not until the 1930's that the concept of "sanitary landfilling" was first proposed in describing a cut-and-cover operation in California (Gendebien *et al*, 1992). As the volume and hazardous nature of waste arisings has increased, so the potential for environmental pollution has increased, with in some cases serious consequences. This has led to an increased awareness of the potential environmental impact of landfilling activities, and a need for stricter controls. Yet even today within Europe, the number of uncontrolled landfills in Italy and Portugal exceeds 60% of the total number of sites, while in Greece and Spain the number of uncontrolled sites represent approximately 30% of the total. In Greece, where 100% of MSW arisings are landfilled, most is disposed in open dumps (Gendebien *et al*, 1992).

In any waste treatment process, there will always be a waste stream that requires disposal to land: the challenge is one of ensuring that the risks associated with landfill disposal are recognised and are acceptable. In both developed and developing countries, but especially in the latter, problems associated with waste disposal could be anticipated to increase, either through increased populations, or as a result of increased industrialisation. In this case, it will be important that the knowledge and systems from other countries are effectively and **appropriately** transferred to ensure that the same mistakes are not repeated. As long as landfills are well designed, well operated, and the hazards are understood, they are likely to continue to be an important and cost-effective means of managing waste.

Although factors such as those above will increase the volume of waste arisings, it seems certain, having recognised their potential for both economic and environmental savings, that waste minimisation and recycling are likely to become increasingly important, and will reduce the total amount of waste for disposal. The net effect will depend upon the balance of the opposing forces, but in many countries, changes in the composition of waste arisings could be anticipated; the incineration of waste is likely to increase, and alter the nature of waste for disposal. However, matter can neither be created nor destroyed, and when burning waste in incinerators, much of the material that would otherwise have been disposed to landfill, (and which would ultimately be dispersed to land), is dispersed as carbon dioxide, and a range of other gases, to the atmosphere. For the remaining ash, landfill disposal remains one of the few suitable options (this issue is discussed further in Chapter 7).

The increased awareness of the problems associated with landfilling has resulted in a great deal of research and development activity, and an increase in the amount of landfill related legislation. Around the world, the impacts of new legislation are being assessed, further legislation is expected, and the relative merits and most suitable means of operating and controlling landfills are undergoing continual debate. However, whatever the future may hold, the environmental impact will be a major consideration in determining the direction of future landfilling activities. Recognition of the pollution potential of landfills has led to the development of new concepts for landfill design and operation. The sustainable landfill is one such concept and is based upon the requirements of sustainable development as defined at the The Earth Summit '92 (The United Nations Conference on Environment and Development, held in Rio de Janeiro in 1992). The requirements of sustainable development are likely to have a major influence on landfill design and management in the future, and is therefore worthy of further consideration at this point.

1.2   Sustainable development

"Sustainable development" has been defined in 1987, in the report of the World Commission on Environment and Development (The Brundtland Report) as "development which meets the needs of the present without compromising the ability of future generations to meet their own needs". This report identified the characteristics of sustainable development as:

* The maintenance of the overall quality of life
* The maintenance of continuing access to natural resources and
* The avoidance of lasting environmental damage.

These concepts are also iterated in the 5th European Action Programme (CEC, 1992a) which states that "sustainable development implies putting in place a policy and strategy for continued economic and social development without detriment to the environment and the natural resources on which human activity depends". The EC's view on inputs to effective sustainable development are summarised in Figure 1.

The concept of sustainable development recognises that both economic and environmental factors affect 'quality of life', and that any new development must account for the **environmental** cost as well as the **economic** cost. For a landfill development, the environmental cost will include, for example, the cost of leachate treatment to a level that will be sustainable in the long term. i.e. it will not compromise the inherent value and integrity of the environment for future generations. "Sustainability" has also been defined (Ruckelshaus, 1989), "as the nascent doctrine that economic growth and development must take place, and be maintained over time, within the limits set by ecology in the broadest sense - by the interrelations of humanbeings and their works, the biosphere and the physical and chemical laws that govern it". It is worth noting that in this definition growth and development **must** take place, and therefore "sustainable" (or derivations thereof) should always be closely linked with "development". i.e. we must continue to grow and develop in a way determined by the robustness (or the sensitivity) of the ecosystem. Ruckelshaus goes on to say that environmental protection and

- Air quality management
- Water resources management
- Soil quality maintenance
- Nature, landscape, conservation
- Energy security and efficiency
- Demographic management (incl.urban env., public health and safety)
- Waste management

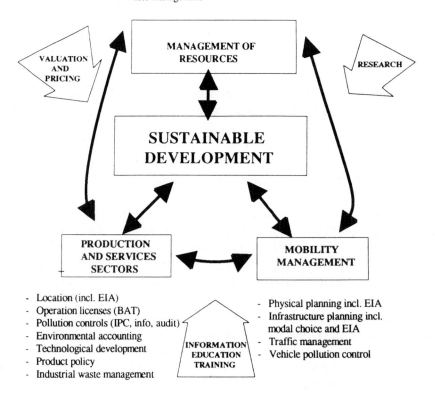

- Location (incl. EIA)
- Operation licenses (BAT)
- Pollution controls (IPC, info, audit)
- Environmental accounting
- Technological development
- Product policy
- Industrial waste management

- Physical planning incl. EIA
- Infrastructure planning incl. modal choice and EIA
- Traffic management
- Vehicle pollution control

Figure 1   Inputs to Sustainable Development

With Permission of European Communities (CEC, 1992)

economic development are complementary rather than antagonistic processes - a view that until relatively recently, few within the industrial sector would have supported. Even now many will disagree with the above sentiments, - an attitude born out of ignorance of the long-term effects of mankind's activities, and/or because of an attitude of short-termism, especially when related to finance.

The European 5th Action Programme (CEC, 1992a) emphasises that issues such as climate change, air and water pollution and waste related issues are only **symptoms** of mismanagement and abuse, and that the real problems are the current patterns of human consumption and behaviour. This emphasises the need, not only to manage the environment more effectively at an institutional level, but also the need to educate and involve the general public in sustainable activities.

The Earth is not infinite in its resources, nor in its capacity to absorb the effects of highly polluting activities, and for ensured survival of the Earth and everything in or on it, these non-sustainable practices must be replaced with those that are sustainable and which will allow future generations to use, enjoy, and benefit from the Earth's environment in the same way that we have been able to. The attainment of sustainable waste management, and the development of sustainable landfilling practises will be an important step in achieving this.

The overfishing of the Earth's oceans is an example of a topical non-sustainable practice; in the North Sea, the numbers of Herring and North West Atlantic Haddock are down by 75% and 90% respectively (Kersey, 1994). These decreases are a consequence largely of over-fishing by European fishing fleets. Such activity is clearly far from sustainable, and is based solely on short-term economic gains. Another topical issue is, at the time of writing, occurring at the follow - up summit to the Earth Summit '92 which is taking place in Berlin, and where the major issue is that of control of carbon dioxide emissions to atmosphere. Here, discussions are taking place in the light of concern that global warming is causing a breakdown of the polar ice caps. Whereas in the past, it has not been possible to categorically state that global warming is a real phenomenon, many scientists now feel this to be the case. If proved correct, then global warming, caused by the release of various gases to the atmosphere, is a clear example of development that is far from sustainable and where the **environmental** effects of mankind's activities have been ignored. The relationship between landfill and global warming is discussed in Chapter 4.

For waste management, The European Commission states that "Despite Directives going back to 1975 on waste in general, on toxic and hazardous waste and on transfrontier shipment of waste, management of the Community's enormous waste stream is far from being under control" (CEC, 1992a) The same source identifies a 13% increase in municipal waste arisings over the last 5 years, despite increased recycling of paper, glass and plastics. Thus there is much to be done to implement effective and sustainable waste management even in a region with the second highest per capita GNP.

The European Commission effectively summarised the current position when it stated (CEC, 1992a) that "one of the major shortcomings of economic policy in the past has been its failure to take into account or measure accurately the full costs imposed on the environment. Historically, the Earth's ecosystem has been treated as an infinite source of raw materials, energy, water etc. .... Failure to properly account for, cost and value the environment and environment policy may lead to a wholly misleading understanding of society's wealth, it's income and it's real sustainable development potential. Policies which are designed to promote economic development are doomed to eventual failure if they do not include the environmental dimension as an integral component".

Many changes will be required for this to occur, including changes in societies' culture to one more 'in tune' with sustainable development; more effective use, and re-use of materials and energy, and the incorporation of the **environmental** 'cost' and not just the **production** cost of a product within the selling price. The environmental 'costs' are often referred to as the "externalities", and should account for any actions required to ensure sustainable production. For waste disposal, the externalities have been identified as comprising the sum of site disamenity costs, global pollution costs, conventional air pollution costs, air toxics, leachate costs, and transport-related costs, minus displaced pollution damage (DoE, 1993a). Each of the above are applicable to landfill, and would include, amongst other costs, the cost of leachate collection and treatment to acceptable environmental standards, and any closure and aftercare requirements, pollution control measures, and the cost of any environmental repairs. The wholesale change in attitude and culture required in both the general public and industry to achieve the above will be difficult to encourage and develop, but there is an increasing awareness of the problems described, at both a scientific and policy-making level, and the first steps have been taken to try to address the key issues, and to move towards more sustainable management of The Earth and it's resources.

This increased awareness is evidenced, for example, by current European and US environmental policy, and by meetings such as The Earth Summit '92. Included amongst the achievements of the Earth Summit in 1992, were the Rio Declaration on Environment and Development (27 principles defining the rights and responsibilities of nations in this area), and Agenda 21 (a blueprint for global actions to support the transition to sustainable development), which identifies the basis for action, objectives, activities, and the means of implementation to support the Rio Declaration.

The Rio Declaration sets out the general principles seen as necessary for effective implementation of sustainable development, and includes:

- The integration of environmental concerns within the development process
- The decrease of disparities in standards of living for all
- The adoption of a precautionary approach to protection of the environment
- The development of legislation to protect the environment, and
- The participation and co-operation of all nations, at all levels, to achieve the above

Chapter 21 of Agenda 21 specifically addresses "Environmentally sound management of solid waste and sewage-related issues" and was written in response to the conference assembly's decision that enviromentally sound management of waste was among the environmental issues of major concern in maintaining the quality of the Earth's environment, and especially in achieving environmentally sound and sustainable development in all countries. Agenda 21 states that environmentally sound waste management must go beyond the mere safe disposal or recovery of waste that is generated and seek to address the root cause of the problem by attempting to change unsustainable patterns of production and consumption. Four major waste-related programme areas were identified. These were:

- Minimising waste
- Maximising environmentally sound waste re-use and recycling
- Promoting environmentally sound waste disposal and treatment and
- Extending waste coverage service

Within these four programmes, objectives that relate particularly to landfill include:

- To stabilise or reduce the production of waste destined for final disposal, over an agreed time-frame, by formulating goals based on waste weight, volume and composition and to induce separation to facilitate waste recycling and re-use
- To strengthen and increase national waste re-use and recycling systems

For national governments ("according to their capacities and available resources and with the co-operation of the United Nations and other relevant organisations, as appropriate"), the following objectives apply:

- By the year 2000, establish waste treatment and disposal quality criteria, objectives and standards based on the nature and assimilative capacity of the receiving environment
- By the year 2000, establish sufficient capacity to undertake waste-related pollution impact monitoring and conduct regular surveillance, including epidemiological surveillance, where appropriate
- By the year 2025, dispose of all sewage, waste waters and solid waste in conformity with national or international environmental quality guidelines

Thus for waste management, and landfill disposal, the objectives are clear and apparently achievable, and in this way, Agenda 21 represents a step towards the attainment of common standards across the world, (in those aspects of landfill management that are specifically referred to), and which helps to ensure progress towards sustainable waste management. Whether or not landfill is truly sustainable is a moot point, and one which will be considered throughout this book. The concept of 'acceptable risk' and the "nature and assimilative capacity of the receiving environment" are important factors in this discussion and are considered later.

Within Europe, The Maastricht Treaty introduced as a principle objective, the promotion of sustainable growth with respect to the environment. The word "sustainable", as used here is intended "to reflect a policy and strategy for continued economic and social development without detriment to the environment and the natural resources on the quality of which continued human activity and further development depend". The treaty specifies that this policy "must aim at a high level of protection and that environmental protection requirements must be integrated into the definition and implementation of other community policies".

However, implementation is reliant upon the co-operation of national governments and industry. There is some evidence that the first steps have been taken by elements of each, but the culture shift required for much of industry to move to a truly sustainable method of operation is immense and will take many years to achieve, if indeed it is ever achieved. The 'conversion' of industry may be encouraged, and hastened by national governments (e.g. through policy and legislative/economic control), and by social pressure (pressure by the public and by pressure groups), but at the moment, too many governments work to short term policies aimed at ensuring re-election or economic growth (at any cost), rather than at more long-term policies required for encouragement of sustainable development. Yet in many cases where sustainable policies have been implemented, there has also been financial gain, (e.g. through waste minimisation gains), and within the waste industry, there is considerable potential for financial gain as a result of following sustainable policies. These financial gains may arise in a number of ways, including reduction of waste disposal costs (as a result of waste recycling and minimisation), reduced long-term environmental liabilities and decreased insurance premiums.

Legislation in the USA (such as RCRA) and from Europe - such as the proposed European Landfill Directive (CEC, 1991) encourages a more sustainable landfill practice, although does not prevent all non-sustainable elements of landfill disposal. However, for effective and sustainable waste management, landfill cannot be considered in isolation but must be viewed within the context of integrated waste management. Recently developed strategies for waste management, such as those in Europe and the USA, clearly show the importance of such an holistic view, and the increased attention that is now being given to sustainable waste and landfill management.

1.3   Landfill as part of an integrated waste management strategy

European policy on waste management serves as an example of integrated waste management policy. The 5th Environment Action Programme (a programme which sets out a strategy and timetable for community actions on the environment), 'Towards Sustainability' (CEC, 1992a) sets long-term policy objectives and intermediate targets for the year 2000. This 5th action programme differs from the previous four in that it is more pro-active, incorporates socio-economic, as well as environmental goals, and aims to involve all sectors of the community in achieving the aims of the programme.

Sustainable waste management forms the basis of the 5th Environment Action Programme, and for municipal waste, the overall target is the "rational and sustainable use of resources", achieved through a hierarchy of management options. These are:

- Prevention of waste
- Recycling and re-use
- Safe disposal of remaining waste in the following ranking order
  - Combustion as fuel
  - Incineration
  - Landfill

Policy in the USA is very similar to that in Europe, and is based upon a hierarchy of waste management options. The Resource Conservation and Recovery Act (RCRA) was introduced in 1976 and provided for what is known as "cradle to grave" control of waste - from its generation to final disposal. Amendments introduced in 1984 (known as the Hazardous and Solid Waste Amendments or HSWA) introduced further controls where, for example, much of the untreated hazardous waste arisings were banned from landfill (effective within a specified time), and where waste generators were required to introduce a waste minimisation programme. In 1989, the US Environmental Protection Agency (USEPA) produced "The Solid Waste Dilemma: An agenda for action", which has been continually updated since that time. A wide range of waste management issues are addressed within this report, which is based upon the principle of integrated waste management, where an appropriate combination of source reduction (waste minimisation, but also defined to include increase in the useful life of products), recycling, combustion and landfill are used to manage waste in an environmentally safe manner. As in European policy, these options are arranged in a waste management hierarchy with source reduction being the preferred option, followed by recycling (defined to include composting), incineration and landfill.

Sub-title D of RCRA relates to solid waste management and requires, amongst other things, regional long - term planning and co-operation between federal, state, and local planning to assist in developing and encouraging environmentally-sound methods for the disposal of solid waste, maximisation of the utilisation of valuable resources within the waste (including energy), and encouragement of resource conservation. Planning is required to account for future needs and should account for the impact of waste minimisation and recycling initiatives. The specific requirements for landfill design and operation will be considered later.

The effective implementation of a policy requires a strategy, and within Europe, a number of countries including the Netherlands, Germany, and the UK have now published national waste strategies which take account of the above waste hierarchy. The most recent was produced as a draft consultation document, in January 1995 by the UK Government. It was entitled "A Waste Strategy for England and Wales" (DoE, 1995) and presented the UK Governments' interpretation of the European Commission waste hierarchy for waste management in which sustainable waste management is used as a basis for the new waste strategy. This draft strategy proposes a "five point plan" to achieve the overall aims. The five strategies adopted are - those based upon regulatory control, market based factors, planning control, promotion, and information. The market-based strategy is one that has received much attention within the UK, and a number of studies have been undertaken on behalf of the Government. "Landfill costs and prices: Correcting possible market distortions" (Coopers and Lybrand, 1993) examined the costs of landfill, incineration, and recycling activities and assessed the potential affect of a landfill levy on the disposal of waste. "Externalities from landfill and incineration" (DoE, 1993a) provided a comparative evaluation of the environmental impacts of landfilling and incinerating waste.

The market-based approach can be seen in the UK Government's budget statement of 1994, which announced a new landfill tax. The draft waste strategy for England and Wales (DoE, 1995a) states that "the underlying purpose of the new tax is to ensure that landfill waste disposal is properly priced to reflect it's environmental impact. In turn, this will encourage business and consumers, in the most cost effective and least regulatory manner, to produce less waste; to recover value from more of the waste that is produced, for example through recycling; and to dispose of less waste in landfill sites. By applying the "polluter pays" principle, it will complement and reinforce other policy initiatives to minimise and make better use of waste". In this way, the levy represents a national strategy aimed, in principle, at promoting more sustainable waste management; cynics would argue that the main aim of the levy is to increase Government Treasury funds!

The "polluter pays" principle highlights a market-based approach to waste management, as it seeks to ensure that those directly responsible for pollution are made to pay for that pollution, and forms the basis of recent UK waste management legislation as enshrined in the Environmental Protection Act (1990). In the context of landfill management, this principle is evidenced, for example, by the fees charged to landfill operators by the local Waste Regulation Authority (WRA) for site inspections to ensure conformance with site license conditions, and by the payment of fees to discharge leachate to foul sewer under conditions where the fees charged relate to the pollution potential of the leachate.

The above strategy defined disposal as "the treating or depositing of waste without generating any extra value or benefit from them.". It goes on to say that "It is the lowest level of the hierarchy, and should be the last option to be considered when deciding how to manage a waste stream".

For the UK, the draft waste strategy requires that:

- the distribution of waste disposal facilities must continue to match the demands of sustainable economic development; and
- pollution from individual facilities must be controlled to conform with the requirements of sustainable environmental management, now and in the future

As in other national waste strategies, specific targets have been set and for the UK (in relation to landfill) these are:

- To reduce the proportion of controlled waste going to landfill by 10% by the year 2005, and to make a similar reduction over the next 10 years

An identified National Waste management strategy facilitates the production of local strategies and again using the UK as an example, the draft waste management plan for Greater London (1995 - 2015) was produced in January 1995 (London Waste Regulation Authority, 1995). The production of such a plan is a legal requirement under the EPA (1990), which requires local authorities to produce arrangements for a wide range of waste issues including:

- the treating and disposing controlled waste within it's area of control so as to prevent or minimise pollution of the environment or harm to human health
- the nature and quantities of waste that the authority expects to arise, be imported/exported, or diposed
- the means of waste treatment or disposal within it's area
- the cost of waste treatment or disposal

This more local plan is also set in the context of the waste management hierarchy. The document states that for London, even if all waste was incinerated, there would be insufficient landfill capacity for the remaining ash, and therefore clearly indicates the urgent need to for others to adopt a more strategic approach to waste, and landfill management.

The major difference between the UK draft strategy and the European waste hierarchy is that in the former, composting processes and the recovery of energy from either incineration of waste or from landfill gas, are considered under a 'recovery' category. The UK waste hierarchy (in descending order of preference) is reduction, reuse, recovery, disposal. 'Disposal' is described as "The least attractive option because no benefit is obtained from the materials. The emphasis here must be on ensuring that disposal is environmentally sound". Irrespective of the terms used, it is clear that whether at a European level or at a national level, simple relatively uncontrolled disposal to landfill is, if not altogether a thing of the past, certainly on it's way out.

A further key element in the implementation of a successful waste management programme is the setting of objectives. The USEPA document "The Solid Waste Dilemma: An agenda for action" identified six objectives for waste management. They were:

- An increase in the waste planning and management information available to interested elements including states, local communities, waste handlers, and industry, and to increase data for research and development
- Increase effective planning by states, local communities and waste handlers
- Increase source reduction by industry, government and the community
- Increase recycling by industry, government and the community
- Reduce risks from MSW incineration to protect human health and the environment, and
- Reduce risks from landfill to protect human health and the environment

These objectives are very similar to those identified within the European 5th action plan (Tables 1 and 2) and indicate consensus on what comprises the major problems and how best to deal with the problems of waste management.

Before focusing on landfill activities alone, it is worth looking briefly at the other elements of a waste management hierarchy in order to be able to better identify the role of landfill disposal within an integrated waste management strategy.

1.3.1   Waste minimisation

"We have learned the inherent limitations of treating and burying wastes. A problem solved in one part of the environment may become a new problem in another part. We must curtail pollution closer to it's point of origin so that it is not transferred from place to place".

WILLIAM REILLY
ADMINISTRATOR United States Environmental Protection Agency
DECEMBER 1990 (USEPA, 1990a)

Waste minimisation is not a new concept, but is one that has been practiced in the USA, Europe, and elsewhere for many years (even if only by a relatively enlightened few, in global terms).   A comprehensive appraisal of waste minimisation in the USA, including the legislative background, techniques, assessments and case studies has been undertaken by Freeman (1990 and 1995). However, sustainable waste management initiatives have served to raise the profile of this option, which now forms a key component of national waste management strategies across the world. Clearly, if the production of waste is reduced, then the potential for long-term environmental damage is reduced accordingly. Furthermore, for many industrial and other processes, waste minimisation is achieved through process modification for enhanced efficiency, often with resultant reduction in raw material utilisation and energy costs. The economic savings associated with such reductions are, in most cases, far greater than the savings associated with reduced waste disposal costs, but may never have been recognised without the initial consideration of waste minimisation. Thus while pursuance of sustainable waste management practices will, in many cases, involve short to medium term costs, that of waste minimisation can often lead to short and medium term financial gains. In this way, waste minimisation can provide an ideal vehicle to encourage waste producers to develop a more responsible attitude to the environment and to adopt sustainable production processes.

Table 1   EC Fifth Environment Action Plan - Initiatives and targets related to waste

	Objectives	EC Targets up to 2000	Actions	Time Frame
Municipal Waste	• Overall target: rational and sustainable use of resources			
	• Prevention of waste (closing cycles) • Maximising recycling and reuse of material • safe disposal of any waste which cannot be reused in following ranking order: ⇒ combustion as fuel ⇒ incineration ⇒ landfill	• waste management plans in member states • stabilisation of quantities of waste generated at EC average 300kg/capita (1985 level) on a country by country basis no excedance of 300kg/capita • recycling/reuse of paper, glass and plastics of at least 50% (EC average) • Community-wide infrastructure, for safe collection, separation and disposal • no export outside EC for final disposal • recycling/reuse of consumer products • market for recycled materials • considerable reduction of dioxin emissions (90% reduction on 1985 levels by 2005)	• landfill Directive operational • directive on packaging operational • cleaner technologies and product design • policy on priority waste streams, stop on landfill for specific wastes (legislation + voluntary agreements • reliable EC data on waste generated, collected and disposed • system of liability in place • economic incentives and instruments (include. deposit return systems + voluntary agreements • standards for dioxin emissions from municipal waste incineration	before 1995 1995 progressive ongoing 1995 2000 ongoing before 1994
Hazardous Waste	• prevention of waste (closing cycles) • maximum reuse/recycling of materials • safe disposal of any waste which cannot be reused in following ranking order: ⇒ combustion as fuel ⇒ incineration ⇒ landfill	• no export outside EC for final disposal • waste management plans set up in member states • EC-wide infrastructure for safe collection, separation and disposal • market for recycled materials	• landfill Directive operational • Directive on incineration of hazardous waste operational • policy on priority waste streams, stop on landfill for specific waste • cleaner technologies • reliable EC data on waste generated, collected and disposed • setting up of bourse de dechets • system of liability in place • inventory risks • economic incentives and instruments, include. vol. agreements	before 1995 1995 ongoing ongoing 1995 before 1995 2000 1995 ongoing

Reproduced with Permission of The European Community (CEC,1992)

Table 2.   Strategic chart for a community management policy on hazardous and other wastes

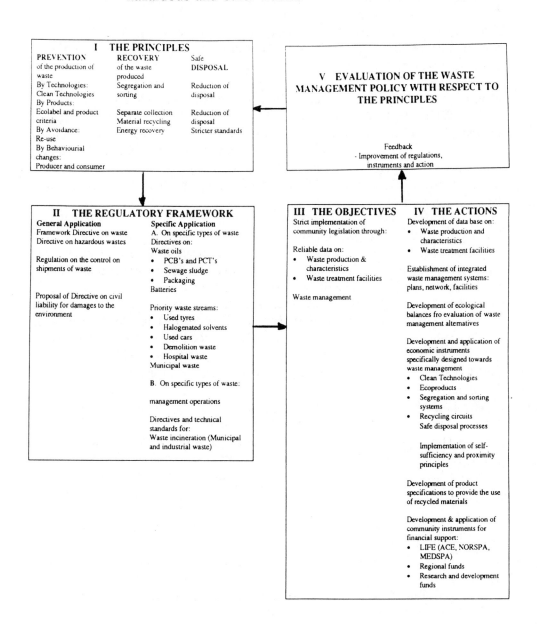

Reproduced with permission of European Communities (CEC, 1992)

As in Europe, waste minimisation is a key element of US policy, endorsed by regulatory requirements including the Pollution Prevention Act (1990), The Clean Air Act(1967) and the amendments of 1990, the Toxics Substances Control Act (1976), and The Resource Conservation and Recovery Act (1976) The Pollution Prevention Act (1990) arose out of growing national concern with waste generation and management practices (Thurber and Sherman, 1995), and In it's deliberations, US Congress declared that "source reduction is fundamentally different and more desirable than waste management and pollution control".

Within the USA, waste minimisation is more often referred to as "pollution prevention", and is very closely allied to the concept of sustainable industrial development. The USEPA defined pollution prevention as "the use of materials, processes, or practices that reduce or eliminate the creation of pollutants or waste at the source. It includes practices that reduce the use of hazardous and non-hazardous materials, energy, water, or other resources as well as those that protect natural resources through efficient use" (cited in Freeman, 1995). US policy is more prescriptive than that in Europe, where any specific waste minimisation requirements have been set at a national level. For example, RCRA places specific requirements upon waste generators to certify that they have a programme in place to reduce the volume or quantity and toxicity of the materials they manage. Such programmes must exist to the extent that they economically practical. In contrast, a recent report showed that almost half of UK industry does not have a waste minimisation policy, and 44% do not know how much they spend on waste (Anon. 1994a). RCRA also establishes as national policy (42USC. 6922(b)) that to the extent feasible, the reduction or elimination of hazardous waste generation should be achieved as expeditiously as possible. Where this cannot be achieved, such waste should be treated, stored, or disposed so as to minimise the present and future threat to human health and the environment.

There are relatively few examples of specific waste minimisation requirements within Europe, where most minimisation schemes remain on a voluntary basis. Of the exceptions, that in Germany is perhaps the best recognised. Here, the aim is to:

- reduce the pollutant content of waste, enabling more recycling of the pollutant-free waste, and
- reduce the amount of household waste by reducing the packaging content

The 1986 Waste Act empowers the Government to:
- subject certain products to mandatory labelling or separate handling
- require manufacturers to reclaim their products once they become waste, and
- impose bans or restrictions on marketing

Although the drive towards waste minimisation in Europe appears to be less advanced than in the USA (where the benefits appear to have been recognised by much of the industrial sector), the importance of this option is becoming increasingly evident; e.g. the Netherlands has now set a target of 10% reduction in the volume of waste produced by the year 2000. In addition, the amount of waste disposed to incineration will be reduced from 24M tonnes to 14.5M tonnes, with a considerable shift away from landfill to incineration. Waste minimisation at the local community level is also increasing, and within the UK some local authorities publish advice to householders on relevant issues, and in some cases, appoint waste minimisation officers (Coggins, 1993). The involvement of the local community is a key element in the adoption of a successful national waste minimisation strategy, for as indicated above, issues such as climate change, air and water pollution and waste-related issues are only **symptoms** of mismanagement and abuse; the real problems are the current patterns of human consumption and behaviour. In the public sector, "waste minimisation" is more likely to equate with source separation, recycling, and recovery.

1.3.2   Waste recycling and re-use

Waste recycling and re-use is, at a policy level, the most favoured option for waste management, after waste minimisation, in both Europe and the USA. The precise definitions vary from country to country and within countries, but in general terms waste recycling is the re-use of materials, after processing (eg the production of "recycled" paper from waste paper), and waste re-use is the re-use of materials after little, if any processing (e.g. the re-use of milk bottles). Waste recycling and re-use is an integral part of life in many poorer, developing countries where the drive to derive maximum use from any particular material is much greater than in the developed world. Examples include the scavenging of waste materials from landfill sites in, for example, South America and India, from which many people derive a living.

However, in the developed world which is relatively rich in resources and where waste disposal (e.g. to landfill) does not account for the associated externalities, and is still relatively cheap, recycling of some materials can be too expensive to maintain unless specifically required through legislation, or supported by national, regional, or local government. In the absence of legal requirements, many recycling schemes that fail, do so solely for economic reasons; these can include high costs of collection and sorting, and changes in the market for the recycled product. However, where there is a will, recycling can be most effective; In Sweden, the amount of waste arisings has been almost the same for 15 years, and of a total household waste arisings of 3.2M tons, approximately 0.5M tons (16%) are recycled, and recovery rates for individual components such as newspapers are high (64%) (Rylander, 1993). If the cost of recycling was compared to the true cost of landfill disposal or incineration at facilities operated in a sustainable manner, then recycling successes such as that in Sweden may be more common. The adoption of a truly sustainable waste management policy would result in a significant increase in the extent of recycling undertaken, and this together with effective waste minimisation, would result in a reduction of waste disposed to either landfill or incinerators.

Despite the the fact that in many cases, waste recycling is currently more expensive than waste disposal, recycling in the USA, for example, has increased from 10% of the solid waste stream in 1987 to 17% in 1994 (Fairbank, 1994). In this case, the driving forces for these increases are social, political and legislative and are based on national and local policy described above. During 1990 in the USA, more than 140 recycling laws were enacted by 38 states, and thirty three states have comprehensive laws which require detailed recycling plans and/or separation of recyclables and which contain at least one other provision to stimulate recycling (Fairbank, 1994).

Similar targets are now being set in Europe, e.g. In the Netherlands it is expected that by the year 2000 90% of building and demolition waste will be recycled (Laurijssens, 1993).

For that waste that cannot be minimised or recycled, combustion is the next most favoured option.

1.3.3 Combustion as a fuel

According to UN projections (CEC, 1992a), the demand for energy will increase from 1990 figures of approximately 9 billion tons of oil equivalent (toe) to around 20 billion toe in 2050. Under the above conditions, the security of energy supply, especially in developing countries, will be placed under considerable stress unless there is a different approach to nuclear energy power, to greater use of waste-related energy sources, or unless there is a breakthrough in the development and use of alternative energy technologies (CEC, 1992a). Projections such as these, and recognition of the detrimental environmental effects of fossil-fuel combustion have led to increased interest in the development of renewable energy sources such as biomass, solar energy and wind energy, and to more effective control of combustion processes.

The recovery of energy from municipal solid waste falls within the category of renewable energy and has received increasing attention in recent years. This has been strongly influenced by a drive to reduce the burning of fossil fuels, thereby reducing the release of pollutants such as gaseous nitrogen and sulphur oxides (NOx and SOx).

The combustion of waste as a fuel is seen by many as a preferable alternative to landfill, where appropriate, and has received much support as a waste **treatment** option. However, according to Wallis and Watson (1995), in general recycling materials saves 2-5 times the amount of energy recoverable by combustion. They cite (anon., 1992a) that even plastics recycling to materials is several times more advantageous than recycling to energy. However recycling is not always feasible (eg for reasons of material contamination, or because of the lack of markets), and there is significant potential for further development of incineration as a major waste management option. Within the UK, the Royal Commission on Environmental Pollution (RCEP) in it's seventeenth report on "Incineration of Waste" (RCEP, 1993) produced a number of recommendations including:

- that the government should give targets to the waste disposal authorities for the recovery of energy from municipal waste and
- that the potential for low-grade waste heat recovery should be reviewed, including the circumstances in which it might be economic in the UK without any specific study.

The report concludes that "incineration of waste at plants constructed and operated to the new standards ought to have an important and growing role in the national strategy". However, incineration is a waste **treatment** option and not a waste disposal option. The ash produced should be disposed to landfill, and incinerators cannot function in the absence of landfill, although the design and management of landfills accepting only incinerator ash will be quite different to those accepting untreated waste. These issues will be explored in more detail in Chapter 7.

The RCEP 17th report also recommended that the UK Department of the Environment (UK DoE) press forward with studies to determine the Best Practicable Environmental Option (BPEO) for particular waste streams. This concept is important modern waste management and is worthy of further consideration at this point.

BPEO was first defined in the UK in the fifth report of the RCEP (RCEP, 1976) as "the optimum combination of available methods of disposal so as to limit damage to the environment to the greatest extent achievable for a reasonable and acceptable total combined cost to industry and to the public purse". In it's twelfth report the RCEP (1988) emphasised that the BPEO is:

"the outcome of a systematic consultative and decision making procedure which emphasises the protection of the environment across land, air and water. The BPEO procedure establishes, for a given set of objectives, the option that provides the most benefit or least damage to the environment as a whole, at acceptable cost, in the long term as well as in the short term".

Thus for a particular waste stream, disposal according to BPEO would be the least environmentally damaging and would therefore constitute the most sustainable waste disposal option. The RCEP twelfth report (RCEP, 1988) outlines the steps required in selecting a BPEO. Because BPEO considers emissions to all media, the total pollutant load on the environment can be assessed. Thus while the pollution potential of incinerator ash within a landfill may be less than that of untreated waste, the BPEO would also account for the potential affect, for example, of the release of material to air. This type of assessment would facilitate comparison of the pollution potential of carbon within untreated waste upon release in landfill leachate (as BOD/COD) and landfill gas, with release of the same carbon as carbon dioxide within the incinerator gas emissions. After making a number of key assumptions, the RCEP 17th report (1993) showed (Figure 2) that the emission of greenhouse gases from landfill were greater than those from incineration. However, these findings have been questioned, and both this and other relevant issues are discussed in more detail in Chapter 7.

Although the technical arguments for incineration are strong, there is much public opposition both within the UK and the USA (as well as elsewhere) to the construction of incinerators in local areas - a phenomenon known as the NIMBY (Not In My Back Yard) syndrome. This phenomenon also applies to landfill developments, but in this case, the level of public opposition is generally less noticeable. Where the public have been more willing to accept incineration, such as in Sweden, Germany and Denmark, this has been ascribed to the development of tighter emission standards imposed in these countries in the 1980's (van Santen, 1993), although the greater emphasis placed upon the utilisation of small-scale energy from waste incinerators that are used to supply district heating, is also likely to influence perception in countries such as Sweden and Denmark.

Some statistics relevant to waste incineration are given below (RCEP, 1993):

- The former West Germany has 47 municipal waste incinerators, most of which burn 200,000 - 300,000 t.p.a. Sweden has 23 municipal waste incinerators (treating just over half of municipal waste arisings in 1990) (ISWA, 1991); in the USA there were about 170 in 1991, and in Japan there are about 1900 municipal waste incinerators where nearly 75% of municipal waste arisings are treated.

- In the Netherlands, approximately 40% of municipal waste (about 3 million tonnes) is incinerated (one third is landfilled and one third incinerated) (ISWA, 1991) in incinerators most of which were commissioned in the early 1970's. This figure is expected to rise to 5.3 million tonnes by the year 2000.

• In the former West Germany, energy recovery from incineration reduces the cost of waste disposal by 20% - 40%, in Sweden all municipal waste incinerators recover energy (ISWA, 1991), and in 1991, 137 incinerators in the USA (where approximately one sixth of municipal waste is incinerated) recovered energy.

Progression towards more sustainable waste management will lead to further increases in incineration of municipal and other waste and is likely to affect the design and operation of landfill in the future, and although some concern has been expressed about reduction in the calorific value of waste as a result of recycling of high CV materials (such as paper and plastic), experience from Denmark where CHP incinerators are common and where there are plans to to stop the landfilling of degradable waste by January 1997 (Haukohl, 1993), suggests that the steadily increasing per capita waste production more than offsets any recycling effects.

Thus, while there are a number of difficulties associated with the incineration of waste, the long-term prospects for this waste treatment process appear to be good and will impact upon both the medium- and long-term future of landfill. The incineration of refuse-derived fuel seems less certain.

1.3.4   Refuse-derived fuel

Refuse-derived fuel (RDF) is made by refining municipal solid waste in a series of mechanical sorting and shredding stages to separate the combustible portion of the waste. Either a loose fuel, known as fluff, floc or coarse RDF (c-RDF), or a densified pellet or briquette (d-RDF) is produced (Anon., 1993a). Early development of the process occurred in the UK and Italy where there are a number of RDF plants. Other plants are used throughout Europe (including Germany, the Netherlands and Switzerland), and in the USA. As a general rule d-RDF is easier to handle, store, and transport, although c-RDF requires less refining and processing and can avoid the need for drying the product (a potential source of odour). The costs associated with c-RDF are reduced correspondingly. As with other incineration processes, the ash product must be disposed elsewhere (often to landfill) and cannot operate independently of other waste management facilities. Furthermore, a large percentage of input waste may be rejected [e.g. around 60% of input material at Byker RDF plant, Newcastle, UK is rejected (Anon., 1993a) and will also require alternative means of disposal. When incineration with energy recovery cannot be undertaken, incineration alone is preferred to landfill in the hierarchy described above. Thus, it is clear that for sustainable waste management, as encouraged through a waste management hierarchy, landfill represents the final waste disposal option. This view has been translated into potentially increasingly stringent controls over landfill as identified within the draft Council Directive on the landfill of waste [COM(93)275] (CEC, 1991), and within legislation elsewhere, such as that enshrined within RCRA.

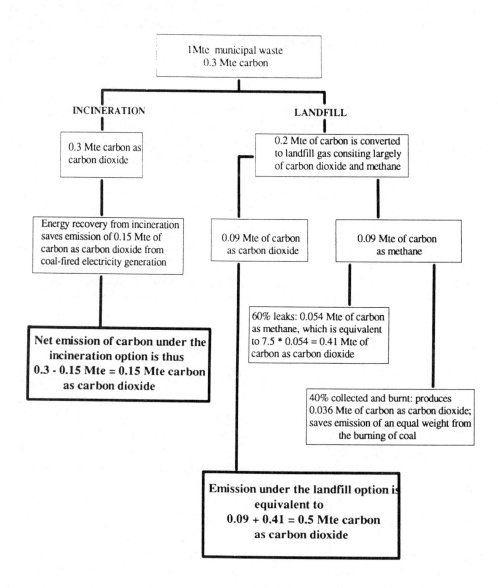

Figure 2.   Incineration versus landfill: Emissions of greenhouse gases
            (RCEP, 1993)

Reproduced with permission

1.3.5   Composting

Composting is an **aerobic** microbial process which degrades organic matter to produce a **relatively** stabilised residue; carbon dioxide is the main gaseous product. It is not a new process, and most gardeners will be aware of the benefits of composting. It can be used to treat a number of waste streams, as long as they are essentially organic in nature, and as well as reducing the pollution potential, the process is effective in the sanitisation of waste as a result of the high temperatures that are reached. The optimal temperature for composting is in the range 45-55^{0}C, and temperatures well in excess of 60^{0}C can be attained (Anon., 1993b). Composting does not, at the current time, constitute a major waste management option for MSW, but has potential to act as a pre-treatment option prior to landfill or as a process for the production of a soil conditioner. In the former case, pre-sorting of waste will not be necessary, but in the latter, it will be essential if there is to be no concern about the quality and safety of the final product. Such pre-sorting will be necessary to remove glass and other sharps, and potentially dangerous chemicals that may be found in the starting material. Bardos (1992) has shown that if source separation is not practiced then the chances of removing the above contaminants by mechanical sorting are extremely remote. Because of this, large scale-production of a marketable product will always be difficult. i.e. Few, if any, markets will accept a contaminated product - even if the contamination level is very low; the degree of control and assurance will always be a problem in source separated organic material.

There are three basic methods of composting - windrowing, the aerated static pile, and anaerobic digestion followed by composting of the residue (Anon., 1993b). For each method the key factors are that there should be sufficient moisture, and that there should be effective distribution of air. Disadvantages include the requirement for relatively large areas on which to place the material to be composted, and unless special precautions are taken, the potential for nuisance due to odour, vermin, insects and birds. For example, although the composting industry in the USA is relatively well developed, a number of units have closed down because of odour-related problems (Anon., 1993b).

In a study undertaken in the UK (Wheeler et al, 1994) the major conclusions included that:

- source separation of municipal waste produces products with lower contamination than mechanically separated products
- source separated municipal waste soil improvers are possible and can compete with traditional media in many sectors
- the potential production of municipal waste-derived composts and digestates is 6.9-9.4 million tonnes
- 98% of the market for waste-derived composts and digestates is in the agricultural sector (total market = 173.1 - 443.7 million tonnes).

The conclusions were not all encouraging, and users expressed concerns over a range of factors including cost, safety, appearance and odour. However, composting in continental Europe is more effective than in the UK (Anon., 1993b), and in Germany, the interest in composting is being fuelled by a directive which stipulates that landfills can accept no more than 5% organic material.

One of the countries that has shown the greatest interest is the USA, where legislation in many states has banned yard waste from landfills, with the result that there are an increasing number of yard waste composting facilities (1,407 in 1990 (Anon., 1993b)). In 1993 there were 20 MSW composting facilities (representing significant financial input - as one facility may cost tens of millions of dollars), with waste inputs of between 10 and 1000 tonnes per day. At inputs such as these, composting is unlikely to become a primary treatment option for MSW. However, after minimisation and recycling, and with suitable waste segregation, composting is likely to be able to provide a local solution for the treatment of small volumes of waste, which after processing may be landfilled, or which may provide a marketable product for sale. Issues concerning the landfilling of biologically pre-treated waste are discussed in chapter 7.

1.3.6   Anaerobic digestion

Anaerobic digestion of organic waste has been practiced for many years in sewage treatment works where the process has been well characterised and where a significant degree of control can be exercised. Anaerobic digestion is also the major process in most landfills - producing methane and carbon dioxide as the major gaseous products, and a **relatively** stabilised residue. Energy can be derived from the methane gas, and because generally less energy is required to run an anaerobic process than an aerobic process, the anaerobic digestion of MSW can, at least in this respect, offer a number of advantages over composting. However, within landfills, process control is poor, and the means of improving control is a key factor in the development of the sustainable landfill. For the controlled anaerobic digestion of organic waste such as MSW to be effective, control must be improved.

As in most composting processes, this control can be improved by segregating waste to isolate the easily degradable organic fraction, and by shredding the organic material to increase the surface-area:volume ratio. A number of anaerobic digestion plants have been built, some of which are described briefly below.

- The RefCoM (refuse converted to methane) digested mechanically sorted organic fraction in a stirred tank reactor. The retention time was 6-27 days (Chynoweth and Legrand, 1988)

- The Dranco process (Dry anaerobic composting) (De Baere and Six, 1988) also utilises the segregated organic fraction of MSW. This is mixed with recycled effluent, and fed to the reactor. The retention time is 18 days, and after post-fermentation processing, dewatering, and drying, the final product is marketed
- The VALORGA process (Zaoui, 1988) utilises a full materials segregation plant at the front-end of the system, has a retention time of 15 days, and produces residues that may be marketed as fertiliser

Whether discussing anaerobic digestion or composting processes, one of the major problems will be the fact that each type of plant is capable of accepting only a small fraction of the waste compared to that which can be disposed to landfill. Because of this, they are unlikely to replace landfill as a waste treatment option, but may, in conjunction with effective waste minimisation and recycling, provide an effective solution to local, rather than national waste management problems. Schemes such as those described above have met with only limited success as commercial ventures, but as attitudes to waste management change, they (or similar schemes) may enjoy more success in the future.

Before assessing the potential for landfill to be a sustainable waste management option, the following chapters will consider the pollution potential of landfill, and the associated risk, through examination of the nature of landfill emissions, the means by which those emissions may reach sensitive receptors and the means for controlling these processes.

CHAPTER 2

PRINCIPLES OF LANDFILL PRACTICE

2.1   Introduction

Landfill has been defined (ISWA, 1992) as "The engineered deposit of waste onto and into land in such a way that pollution or harm to the environment is prevented and, through restoration, land provided which may be used for another purpose". Unfortunately, there are many examples of environmental pollution that have arisen as a result of landfill activities, and while the above may define a model landfill, the reality has, all too often, been far from ideal. The hazards associated with different landfills will vary according to the nature of the emplaced waste, the conditions within the landfill, and the nature of the surrounding environment. An understanding of conditions such as these will help to identify potential environmental risks. For example, gas and leachate production will be very different in "old" and "new" landfills. In this context, a "modern" landfill is defined as one which has accepted a high proportion of biodegradable material and which is different to an "old" landfill which has accepted relatively inert material.

The key requirement of a "sustainable" landfill is prevention of harm to the environment - potentially achieved in a number of ways, but which, irrespective of the chosen principle, requires effective control of waste degradation processes and effective landfill design, engineering, and management, so that the long-term environmental risks are acceptable. The principles of landfill practice used to achieve this have changed considerably since the 1970's and three major principles of landfill design and management have been recognised; these are "dilute and attenuate", otherwise known as "dilute and disperse", "containment", and "entombment" (or "dry-tomb").

2.2 **Dilute and attenuate** is the principle of landfill disposal for unconfined sites, with little or no engineering of the site boundary, in which leachate (that liquid formed wihin a landfill site that is comprised of the liquids that enter the site, including rainwater, and the material that is leached from the waste as the infiltrating liquids percolate downwards throught the waste) formed within the waste is allowed to migrate into the surrounding environment. This principle relies upon attenuation of the leachate both within the waste and in the surrounding geology, by biological and physico-chemical processes. Dilution within groundwater further reduces the risk posed by the migrating leachate - but by definition, necessarily contaminates that groundwater. For the dilute and attenuate principle to be effective, the associated risk should be deemed to be acceptable. Such conditions **may** occur, for example, when the groundwater into which the leachate is discharging cannot be used for public supply, and where the risk to other sensitive receptors is also acceptable, or where there are no other sensitive receptors.

On a timescale measured since mankind first deposited waste in specific locations (landfill), most sites have operated on the dilute and attenuate philosophy. In the relatively recent past, this principle has been supported by studies in the mid 1970's (DoE, 1978) which showed attenuation of leachate as it moved through various unsaturated strata. These studies also showed that the attenuation processes (defined to include dilution) could be used to effectively treat landfill leachate as it migrated from the site. The advantages of the dilute and attenuate landfill are that there is no requirement for expensive landfill lining/engineering, and as liquids formed within the site migrate from the base, there is no requirement for leachate collection and treatment facilities. These "advantages" necessarily have pollution implications.

The effectiveness of this principle is dependent to a large extent upon the effectiveness of the degradation of waste within the landfill, and upon the geologic nature of the strata surrounding the landfill. In countries such as the Netherlands where the water-table is very close to the surface, the disposal of waste to dilute and attenuate landfill would be likely to pollute the surrounding environment to a much greater extent than, for example, in countries where groundwater is located in deep aquifers. In this case, relatively impermeable clays can, in some cases, limit migration of leachate to these aquifers and unsaturated strata may facilitate attenuation processes in the migrating leachate. Thus while local geologic conditions within the UK facilitate disposal to landfill, those in the Netherlands have favoured a greater degree of waste treatment via incineration (as identified above).

However, these attenuation processes cannot be relied upon in all circumstances, and although groundwater and surface waters appeared free of contamination by landfill leachate for many years, recent examples (Anon., 1993c; Anon., 1992b) have shown that some groundwater contamination by landfill leachate has occurred. That this has happened is due to the fact that dilute and attenuate landfills have been operated irrespective of local conditions and without prior risk assessment.

In 1980, the introduction of the European "Groundwater Directive" (The Protection of Groundwater against Pollution caused by Certain Dangerous Substances - 80/68/EEC) (CEC, 1980) caused a reasessement of the 'dilute and attenuate' principle. The Groundwater Directive prohibited the direct or indirect discharge into groundwater of List I (most potentially polluting) substances, and limited discharges of List 2 substances unless prior investigation showed that pollution of groundwater would not occur, or unless the groundwater was permanently unsuitable for other purposes. Because many of the substances present in List 1 and List 2 (Table 3) could be found in landfill leachate, it was clear that better control would have to be exercised. In response to duties under the Groundwater Directive, the European member States implemented new legislation; within the UK, the National Rivers Authority (NRA) under the Water Resources Act (1991), issued a groundwater protection policy (NRA, 1992) in which their policy towards the protection of groundwater was identified. Table 4 shows the decision grid used by the NRA when assessing the advisability of new landfill development, and highlights the importance given to groundwater protection. The introduction of this policy within the UK is one of a number of factors that has encouraged a move towards the containment landfill, which is rapidly becoming the only acceptable engineering means of landfill disposal. Elsewhere, legislation (such as RCRA in the USA) is also driving landfill management to that based on the containment principle.

However, a large number of sites that were developed prior to the 1990's, continue to operate on the dilute and attenuate principle, and have, so far, shown no noticeable detriment to the environment. In this case, the hazard is clearly present, and either local conditions are reducing the associated risk, or it is merely a question of time before a sensitive receptor is affected. For some sites with appropriate geology/hydrogeology, and with a suitable, supporting risk assessment, landfills based on the dilute and attenuate principle may still be technically feasible. Indeed, it could be argued that dilute and attenuate principles represent the only truly sustainable option for landfill management: If the long-term containment of waste is considered necessary, then one must assume that the waste-associated risks have been deemed to be too great. However, sustainable landfill development of the future would require that the long-term waste-associated risks were acceptable. As long-term containment cannot be guaranteed, nor effectively maintained, the risk assessment process must assume that containment will ultimately fail, and the key in the development of a sustainable landfill must be hazard reduction. This may be achieved by either pre-treating waste or by managing the landfill as a bioreactor in which the pollution potential

Table 3.   List I and List II substances as defined by EC Groundwater Directive (80/68/EEC)

LIST I OF FAMILIES AND GROUPS OF SUBSTANCES	LIST II OF FAMILIES AND GROUPS OF SUBSTANCES*
List I contains the individual substances which belong to the families and groups of substances specified below, with the exception of those which are considered inappropriate to List I on the basis of a low risk toxicity, persistence and bioaccumulation.	List II contains the individual substances and the categories of substances belonging to the families and groups of substances listed below which could have a harmful effect on groundwater.
Such substances which with regard to toxicity, persistence and bioaccumulation are appropriate to List II are to be classed in List II.	1. The following metalloids and metals and their compounds:
1. Organohalogen compounds and substances which may form such compounds in the aquatic environment.	1 Zinc 2 Copper 3 Nickel 4 Chrome 5 Lead
2. Organophosphorous compounds	6 Selenium 7 Arsenic 8 Antimony
3. Organotin compounds	9 Molybdenum 10 Titanium 11 Tin
4. Substances which possess carcinogenic, mutagenic or teratogenic properties in or via the aquatic environment (1).	12 Barium 13 Beryllium 14 Boron 15 Uranium
5. Mercury and its compounds	16 Vanadium 17 Cobalt 18 Thallium 19 Tellurium
6. Cadmium and its compounds	20 Silver
7. Mineral oils and hydrocarbons	
8. Cyanides	2. Biocides and their derivaties not appearing in List I. 3. Substances which have a deleterious effect on the taste and/or odour of groundwater and compounds liable to cause the formation of such stances in such water and to render it unfit for human consumption 4. Toxic or persistent organic compounds of silicon and substances which may cause the formation of such compounds in water, excluding those which are biologically harmless or are rapidly converted in water into harmless substances 5. Inorganic compounds of phosphorous and elemental phosphorous. 6. Fluorides 7. Ammonia and nitrates (i) Where certain substances in List II are carcinogenic, mutagenic or teratogenic they are included in category 4 of List I.

* Where certain substances in List II are carcinogenic, mutagenic, or teratogenic they are included in category 4 of List I

Table 4. Landfill acceptability matrix used by the NRA (UK)

SITE TYPE	SOURCE PROTECTION		
	I INNER ZONE	II OUTER ZONE	III CATCHMENT ZONE
1. High pollution potential (landfills accapting domestic, commercial and industrial waste either individually or on a co-disposal basis)	Not acceptable	Not acceptable	Only acceptable with engineered containment and operational safeguards
2. Medium pollution potential (landfills accepting construction, demolition industry wastes and similar)	Not acceptable	Acceptable subject to evaluation on a case by case basis and adequate operational safeguards	Acceptable subject to evaluation on a case by case basis and adequate operational safeguards
3. Low pollution potential (landfills accepting inert, uncontaminated waste)	Not normally acceptable	Acceptable only with adequate operational safeguards	Acceptable only with adequate operational safeguards
SITE TYPE	RESOURCE PROTECTION		
	MAJOR AQUIFER	MINOR AQUIFER	NON-AQUIFER
1. High pollution potential (landfills accapting domestic, commercial and industrial waste either individually or on a co-disposal basis)	Only acceptable with engineered containment and operational safeguards	Only acceptable with engineered containment and operational safeguards	Acceptable only with adequate operational safeguards. Engineering measures may be necessary in order to protect surface warters
2. Medium pollution potential (landfills accepting construction, demolition industry wastes and similar)	Acceptable only with adequate operational safeguards	Acceptable only with adequate operational safeguards	Acceptable only with adequate operational safeguards
3. Low pollution potential (landfills accepting inert, uncontaminated waste)	Acceptable	Acceptable	Acceptable

Footnotes:
- This matrix refers specifically to groundwater protection. In case of non-aquifers, operational safeguards will relate mainly to the protection of surface water resources
- Operational safeguards will include, inter alia, appropriate site management, leachate control and management and monitoring controls
- mono disposal of hazardous industrial wastes will only be acceptable in non-aquifer areas
- For sites which accept wastes of mixed pollution potential, the comments relating to the highest category ofm site type will apply.

Source NRA (1992) This matrix is a summary only and must be read in conjunction with the policy statements (NRA, 1992)

Reproduced with permission

of the waste is reduced over time to a level, where upon containment failure the associated risk is considered acceptable. The acceptability of dilute and attenuate landfills has declined in recent years. To an extent this has been politically dictated, but another important reason for this decline is that while in theory the policy is one that requires some attenuation of waste within the landfill (assuming that migration pathways will lead to sensitive receptors) little has been done to optimise these processes. In many cases, the old dilute and attenuate landfills were such because waste dumping to convenient holes was cheap. The convenient holes were not lined because little thought was given to the pollution problems that may occur. At a time when the whole waste management industry has become much more professional, it may be time to look again at the advisability of a new dilute and attenuate approach to landfill in which the associated risks are recognised and controlled.

2.3   Containment Landfill requires a much greater degree of site design, engineering, and management, and exercises a greater degree of control over the hazards associated with the disposal of waste to landfill: it is a requirement of both sub-title D of RCRA in the USA, and of the proposed Landfill Directive [COM(93)275] (CEC, 1991) in Europe. In the developed world, containment landfill is now the accepted means of disposal to land, although the degree of engineering to achieve containment, and the management of water and other parameters varies considerably.

The underlying principle of containment landfill is that liquids (leachate) generated within the waste should not be allowed to migrate beyond the site boundary. The containment of leachate implies, in most cases (see entombment), that liquid (leachate) will collect and will require treatment. This has placed new requirements upon the effective management of landfills, and is discussed further in Chapter 7. A containment site has been defined (ISWA, 1992) as a "landfill site where the rate of release of leachate into the environment is extremely low. Polluting components in waste are retained within such landfills for sufficient time to allow biodegradation and attenuating processes to occur, thus preventing the escape of polluting species at an unacceptable concentration". According to the site engineering, gas migration may also be prevented or significantly reduced. Perhaps the key phrases in the above definition are "rate of release ..... is extremely low" and "at an unacceptable concentration", for this implies that some migration of leachate may occur, but that the associated risk is acceptable. Even the most highly engineered containment landfills must be expected to fail at some time in the future, whereupon leachate will be released. Recognising this, the operation and management of landfills should be undertaken in such a way that any release will be at an **acceptable** concentration. The application of risk assessment to landfill is an area that has received increased attention in recent years, and is worthy of further consideration at this point.

In general terms, a **hazard** is any event which has the potential to be harmful. When a pathway is present by which a sensitive receptor (target) can be harmed by that event, then there will be an associated **risk**. Some of the potential risks associated with landfill are shown in Table 5. The risks are presented in terms of source, pathway, and receptor. Thus the hazardous event, e.g. the release of odour from the landfill surface, can migrate within the air (pathway) to "harm" a sensitive receptor (people) with an associated risk of loss of amenity and nuisance.

The risk assessment process comprises four stages of investigation (In the following, the terms in parentheses are the terms commonly used in the USA) From Petts (1993):

- Hazard identification (indicator chemical selection) - This may include determination of leachate composition
- Hazard analysis (exposure assessment) - the process of determining releases or event probabilities. For landfill leachate, this could include determination of the the routes via which the leachate could reach sensitive receptors (e.g. people, groundwater), and the characteristics of these receptors
- Risk estimation (toxicity assessment) - the process of assessing harm by determining the dose-effect relationship between the pollutant and the receptor
- Risk evaluation (risk characterisation) - assessment of whether the estimated level of risk (above) is acceptable

For each of the above stages, it is easy to predict difficulties when applied to landfill; for example, we do not know the identity of many trace components of landfill gas; modelling emission transport through geological strata (often fractured) is particularly difficult; for many components of landfill gas and leachate, there are no toxicity data, and the determination of what constitiutes an acceptable risk can vary from individual to individual according to their own individual perceptions. However, according to Petts (1993) "while risk assessment often leads to a quantitative estimation of risks, this is not an inherent requirement and in many cases a qualitative assessment may be sufficient". Petts (1993) also points to problems associated with the uncertainty in risk assessment, and it's use by interested parties to bolster their own particular arguments, but goes on to say that "in essence risk assessment simply provides a structured approach for ascertaining the nature and extent of the relationship between cause and effect".

Clearly the process of risk assessment in relation to landfill is not easy, but it is an approach that sits easily alongside sustainable development; for as indicated earlier, sustainability is of relatively little value without "development", and as long

Table 5   Risks of Landfill (Petts, 1993)

Risks	Pathway	Receptors	Risks
Liner failure Leachate leakage	Hydrogeological	Groundwater Potable water supply The public Rivers etc and associated flora and fauna	Pollution of groundwater Loss of a potable supply Public health risk Damage or loss of flora and fauna
Leachate discharge	Sewer	Sewage treatment works	Effects on biological process
Contaminated surface water	Run-off	Soils, flora and fauna, watercourses, the public via ingestion of water	Water pollution Public health risk Damage to or loss of flora
Gas migration	Geological Soils, landfill cap, to air	Buildings and people Flora	Explosion and fire - death or serious injury. Asphyxiation Damage and loss of flora
Dust	Air	People and flora	Health risk
Odour	Air	People	Loss of amenity & nuisance
Exposed wastes	Direct contact	People	Health risk

Reproduced with Permission

as rigorous standards (such as requirements for multiple liner systems) continue to be used, irrespective of local conditions, there is a real danger that unnecessary over-specification may severely restrict development as a result of the imposition of unnecessarily high costs. A risk-assessment approach to landfill design is recommended in the latest draft UK guidance (DoE, 1995b).

2.4   Entombment or the dry-tomb approach to landfill, in which moisture is, as far as possible, excluded from the waste, is implicit, in the USA, under RCRA sub-title 'D' requirements. The principle of the "dry-tomb" approach is that through prevention of liquid infiltration, the waste will remain dry, will not decompose, and will not produce a polluting leachate. In so doing, this approach effectively stores waste in perpetuity (or for as long as the containment system remains intact), in a relatively dry form. Waste **storage** in this way accepts that no attenuation of waste will occur and argues that either:

- storage creates the opportunity for development of new technologies to deal with the stored waste in a more appropriate way at some point in time in the future, or
- that the engineered containment will remain to always ensure that no infiltration of liquids will occur, and thus potentially polluting leachate will not be produced.

According to Lee and Jones-Lee (1993) the typical features of a "dry-tomb" landfill include:

- A liner which is typically a composite liner (chapter 5), the function of which is to prevent leachate escape to the surrounding environment
- A leachate collection and removal system (LCRS)
- A low permeability cover
- Groundwater monitoring wells.

The converse argument to the "dry-tomb" approach is that waste **treatment**, not storage, is the only safe way forward for landfill practice. This view accepts that containment in perpetuity is not possible and that conditions for waste degradation within the landfill should be optimised in order to encourage attenuation processes. i.e. 'contain and attenuate' landfill and not just simple "containment" represents the safest option for the disposal of waste to land. This concept is explored further in Chapter 7.

In the light of the above principles, and before considering the sustainable landfill, and other relevant landfill principles and practice, it is worth considering current requirements of landfill within the USA (as under RCRA, sub-title D) and Europe (as in the proposed Landfill Directive).

Under sub-title D of RCRA, the more significant requirements are that all new landfills and extensions of old landfills must:

with regard to siting:
- avoid fresh and salt water wetlands
- be outside a 100 year flood plain
- avoid sole source aquifers, unstable areas, seismic impact zones and faults
- be located at least 3048 meters from any airport where jet planes land and 1524 meters from any airport where propeller planes land

with regard to liquid control:
- provide a cap for the landfill which will **prevent** infiltration of precipitation. The cap must have a layer of clay, drainage sand, and top soil for supporting grass to prevent erosion. A geomembrane can be used *in lieu* of clay
- provide a composite liner (clay plus synthetic flexible membrane)
- provide a leachate collection system designed to quickly remove leachate without allowing leachate depth (over the liner system) to exceed 30.48cm
- provide appropriate treatment for leachate removed from landfill
- provide a monitoring programme for measuring groundwater near the landfill

with regard to gas control:
- provide a system for recovering landfill gas and prevent it's migration
- monitor gas migration

with regard to long-term liabilities:
- must provide a trust fund sufficient to maintain the integrity of the cap for a period of at least 30 years

The "dry-tomb" approach is implicit in the requirement to "**prevent** infiltration of precipitation" (above), although insofar as there is also a requirement to collect and treat leachate, there is a recognition that some liquid may be present within the waste. It is highly unlikely that any landfill that is built on the containment principle, and that accepts MSW will be free of leachate, as upon loading with further layers of waste, the liquid within the lower layers will be "squeezed out", and as the waste settles, will rise through the emplaced waste. In these circumstances the amount of leachate produced will be dependent upon the nature of the waste received.

The major requirements of the proposed European Landfill Directive are very similar. They require that:

with regard to siting:
- the location of the landfill must take into consideration requirements relating to:
  - the distances from the boundary to various specified receptors (e.g. residential areas)
  - groundwater/coastal water
  - geological/hydrogeological conditions
  - flooding risks

In most cases, an environmental impact assessment will be required to assess the above.

with regard to liquid control:
-      water precipitations entering into the landfill body shoul be **controlled**
-      the disposal of liquid waste will be banned
-      surface- and/or groundwater should be **prevented** from entering into the waste

  contaminated water and leachate should be collected and treated (unless it is established that the leachate does not pose any threat)
-      At least a single **composite** liner of permeability equivalent to, or less than, 10^{-9}m.s^{-1} should be provided
-      leachate and groundwater should be monitored

with regard to gas control:
-      that measures are taken to control the accumulation and migration of landfill gas
-      that gas be collected treated and used in away which minimises environmental and health and safety risks
-      that gas be monitored for as long as a risk is presented

with regard to long-term liabilities:
-      that adequate financial cover be provided to allow maintenance and aftercare of the site

The key points (above) are not exhaustive but show the similarities between the two sets of controls, where, for example, composite liners, leachate collection and treatment, gas collection, and groundwater monitoring are common. The proposals for the European Landfill Directive have not been ratified, and at the time of writing, the future of the Directive remains uncertain. However, irrespective of the fate of the Directive, many of the requirements are already being implemented in many European countries.

With regard to the fundamental principles of landfill management, the key difference between the two systems is that under RCRA the "dry-tomb" approach is implicit, while under the proposed European Landfill Directive, "control" of infiltration is required, rather than "prevention", as under RCRA. However, the advisability of the dry-tomb approach is now being questioned within the USA (e.g. Maurer, 1993; Lee and Jones-Lee, 1993) and is discussed further in Chapter 7.

2.5   Sustainable Landfill For waste management, the sustainable landfill could be crudely interpreted as dealing with today's waste today and not passing it on for future generations to deal with.   For the purposes of this discussion, a sustainable landfill is defined as one in which the associated risks have been measured and determined to be acceptable. With reference to the model described above (Section 2.2), for a receptor to be at risk, there must be a hazard and a pathway along which migration can occur. In most cases we are not able (nor would we wish) to decrease the sensitivity of receptors to the hazard, and at the same time, removal of the receptors from the field of influence of the hazard is neither practicable nor desirable. Once a contaminant (e.g. leachate) has entered a pathway (e.g. aquifer), then experience has shown that subsequent removal is extremely difficult, if at all possible, and expensive. Therefore, for most practical purposes, the sustainable landfill is one in which acceptable risk levels are achieved through hazard reduction and/or prevention/control of escape to the surrounding environment. The extent to which this can be successfully achieved will be explored in the following Chapters.

     With the concept of acceptable and managed risk in mind, the concept of the sustainable landfill (Harris *et al*, 1993) and fail-safe landfill (Loxham, 1993) have been proposed.

     The proposed sustainable landfill strategy (e.g. Harris *et al,* 1993) accomodates both the landfill disposal of untreated waste and the achievement of final storage quality within 30 years. According to Harris *et al* (1993), "The acceptability of landfilling in the future will depend upon its ability to meet the requirements of sustainable development". They cite Swiss guidelines on waste policy since 1986, that requires that each generation manage its waste to a status of final storage quality - defined as "the stage when any emissions to the environment are acceptable without further treatment". The duration of one generation has been interpreted in Switzerland as 30 years, and is consistent with the 30-year post-closure monitoring period of the European Landfill Directive and RCRA sub-title 'D' in the USA.

     For this to occur, waste must be either pre-treated to a state close to final storage quality, or otherwise, stabilisation within the landfill must be accelerated. (i.e. hazard reduction). While the theory is sound, so far little has been done at the scale of an operating landfill to demonstrate the effectiveness of enhancement techniques for waste degradation. Whether because of this uncertainty, or because of more political factors, a number of European countries have opted for pre-treatment as the means of achieving more sustainable landfill disposal. Germany for example, has limited the amount of organic waste disposed to landfill to 5%, although where this is not feasible, the relevant bodies have been given until 2005 to achieve this goal. In the USA, where all hazardous waste is pre-treated before disposal to landfill, there is considerable public opposition to the use of incinerators for MSW pre-treatment, and thus the incineration option is significantly curtailed, and the dry-tomb landfill predominates at this time.

2.6   Fail-safe Landfill The concept of the 'sustainable landfill' is echoed in the philosphy of fail-safe landfill (Loxham, 1993). The fail-safe philosophy argues that whatever the containment system utilised, it will ultimately fail and/or institutional control will cease, and the contents within e.g., leachate, will be released to the environment. It therefore requires that any releases should be such that the risk posed to the environment is acceptable. For this to be the case, waste disposed to landfill must be pre-treated or degradation must be accelerated such that the hazardous nature of the waste and waste products are minimised.

According to Loxham (1993) "there is a trade-off between on the one hand a site selection and design for a maximum integrity and small failure probability, and on the other, a broader choice of site locations and designs with fail-safe engineering built into them. ...... but the large number of chemical and domestic waste sites will favour the latter design option".

Design on failure techniques involves four steps (Loxham, 1993):

1)   failure scenario definition and failure tree analysis or "how will the engineering fail"
2)   future environmental boundary condition definition or "into what sort of environment will the failure take place".
3)   assessing the environmental impact of failure.
4)   establishing the probability of failure."

This assessed risk approach to landfill design and construction has much to commend it and represents a more versatile and effective means of landfill development than the type of approach that is based on fixed guidelines such as those enshrined within sub-title D of RCRA, and within the European Landfill Directive (CEC, 1991). In order to minimise the risks associated with, amongst other things, liner failure, landraising has been considered as an alternative to landfill.

2.7   Landraising (defined here as landfill in which waste is emplaced with the base at ground level rather than within a hole), is currently receiving greater support within scientific and technical circles, although within the UK a number of planning applications for landraising have been turned down recently on the grounds of 'loss of amenity' (Anon., 1992c). However, in the USA landraise schemes are much more common than in the UK, and Europe. In the hazard-pathway-receptor model described above, landraising does not necessarily reduce the hazard, but can control and limit migration of landfill gas and leachate, and thus exerts control over the "pathway" element.

Typical concerns relate to visual impact, noise, and landscaping, and for landraising to be effective, such concerns must be balanced against the potential environmental benefits. Because of the facility for increased control over emissions, e.g. collection and treatment of gas and leachate, landraising may also allow landfill development in areas otherwise considered to be too vulnerable for landfill development. Vulnerable sites include those located above aquifers, while those sited above, for example, clay could be considered as being suitable for development either with or without a complementary lining system. Other potential advantages include:

- greater ease of provision of underlying unsaturated zone
- location in areas other than abandoned mineral workings
- potential to reduce differential edge settlement
- facilitated surface drainage
- technically simpler

Another important factor in landraising is that for a given surface area, significantly more waste may be deposited, and thus the cost per tonne emplaced is reduced. This means that the high capital costs associated with, for example, the installation of multiple liner systems, and high-technology leachate treatment, can be more easily afforded.

2.8   Landfill restoration and aftercare

Upon completion of the active phase of landfill, the site must be restored in such a way that is appropriate to the surroundings, that minimises the risk to the environment, and which preferably has potential to support productive land-use. In the USA, the overall goals of closure and post-closure care are "to minimise the infiltration of water into the landfill and maintain the integrity of the cover during the post-closure period by minimising cover erosion" (USEPA, 1994). These goals place emphasis on the maintenance of the landfill as a "dry-tomb", that is not present in the European Landfill Directive, where the objectives of the post-closure phase are not specified. However, the systems are similar in many other respects, each requiring the monitoring, and control (if necesssary) of gas, leachate and groundwater, although the specific requirements/monitoring frequencies vary. Also, in the USA (USEPA, 1994), there are much more specific controls (eg cap permeabilities and thicknesses are specified) on the engineering of the landfill cap, which is a requirement for all landfills covered by sub-title D. In the draft European Landfill Directive, there is an absolute requirement for a cap only in those sites that accept hazardous waste, and even in these circumstances permeability criteria are not specified. The minimum cover requirement of sub-title D in the USA is shown in Figure 3.

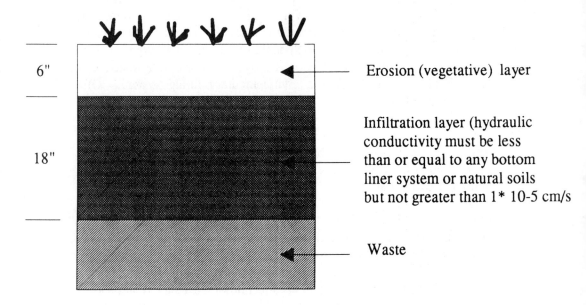

Figure 3.   Minimum requirement for final cover (U.S. EPA, 1992b)

With Permission of USEPA

Landfill restoration is not a "one-off" process, but one that requires monitoring and control for as long as there is an associated risk to the environment. The nature and extent of the risk will be determined by factors such as the type of waste, historical operating procedures, site geology and hydrogeology, and the location and sensitivity of receptors that may be affected by the potential emissions from the restored landfill. There is little scope within this current text to detail the issues of importance in landfill restoration and aftercare. The following identifies some general principles

Effective landfill restoration requires that the restoration planning is conducted at the design stage, so that phasing, materials movement/purchase, costing, etc. can all be undertaken, thus ensuring that the landfill development occurs in the most efficient manner, and which accounts for the many costs associated with this stage of landfill development. The site may be restored to many end-uses, including pasture, cereal farming, public open space, golf courses, and industrial development. The intended end-use of the site will obviously make a big difference, and will be constrained by a number of factors including:

- the level of gaseous emissions
- the nature of the surrounding land, (where for example, for a closed rural landfill, restoration to a farming end-use would be more likely than that to industrial use)
- local topography
- existing flora and fauna (which may be of value, and will need careful managing)
- materials availability (affecting, for example, the type of available top-soil, and the plants, and hence, activities which it will support.

Landscape design is becoming an increasingly important feature of landfill design, and many landfills have been restored such that, the appearance is indistinguishable from the surrounding land. For some time, the question of whether trees were appropriate to landfill restoration was a point of debate, there being some concern over the problems posed by root growth through the capping system, and associated hazards. Such concerns have now been discounted, and there is significant information now available to enable restoration to forestry use (e.g. DoE, 1993b).

If properly planned and managed, then former landfills may be restored in such away that that considerable benefit can be derived. In many cases, the restoration of a former quarry, which may have been a local eyesore, is of benefit in itself.

Aftercare requirements include the maintenance of the restored landfill, and any activities necessary to deal with the potential for environmental pollution. This potential may remain for many hundreds of years according to a number of factors including the nature of the emplaced waste, and the management of the landfill. In a truly sustainable landfill, the aftercare period could relatively small, and so the long-term costs and liabilities would be correspondingly short. Economic factors such as these will be equally as important as legislation in encouraging more sustainable waste management.

The potential costs are enormous and would be required to pay for:

- financial guarantees
- long-term monitoring of landfill gas and leachate
- groundwater monitoring
- leachate collection and treatment
- gas collection and treatment
- settlement monitoring
- any re-engineering of the landfill cap that may be necessary (eg due to settlement or erosion)
- any remediation that may be necessary

This list is not exhaustive, but shows that the sums involved could run into many millions of pounds. To an extent, restoration to end-uses such as golf courses will help to off-set some of these costs, but the most effective way of limiting the long-term costs will be to manage the disposal process in such a way that the associated liabilities are much less and are more predictable.

2.9 Landfill Mining

Landfill mining has been defined (Kornberg *et al,* 1993) as "the excavation and mechanical processing of previously landfilled materials to recover resources (such as land) and to mitigate environmental impacts". The concept of landfill mining is not new; according to Savage and Diaz (1994), a landfill mining project was carried out in Israel in 1953 for the main purpose of reclaiming organic matter for use in agriculture. However, it is only relatively recently that it has been given serious consideration as a waste management option - especially in the USA, where it is a process that has been developed in response to increased difficulties in developing new sites - especially in urban areas, where a number of factors, including public opposition, may prevent development. According to Kornberg *et al* (1993), landfill mining is an attractive proposition for the following reasons:

- landfill mining can extend the life of existing landfill sites and reduce the need for siting new landfills
- it can decrease the area of the landfill requiring closure
- it can remediate an environmental concern by removing a contaminant source
- marketable recyclables can be reclaimed and sold
- energy can be captured through waste combustion

After excavation, mechanical sorting using various screens, classifiers, magnetic separators and conveyors produces a range of output streams that may include soil, ferrous materials, plastics, and aluminium. Of these, soil and the combustible fractions have the greatest potential for re-use - as cover material for landfill and supplemental fuel, respectively (Savage and Diaz 1994). A simple segregation system is shown in Figure 4.

One of the earliest applications of landfill mining and reclamation (LFMR) in the USA was undertaken by Collier County at Naples landfill, Florida, where between 1986 and 1992, the county mined more than 70,000 tons of material, equivalent to 40 - 80 tons per hour, during processing. Approximately 60 % of the material was recovered as a soil fraction (Savage and Diaz 1994). Collier County estimates that recycled waste could produce revenues of up to $35 million over the next decade (Tyson, 1992)

According to Savage and Diaz (1994) the determination of the feasibility of LFMR rests primarily with the site-specific circumstances. Some of the key circumstances are:

- composition and properties of the emplaced waste
- historical operating procedures
- the extent of waste degradation
- the types of markets and uses for the recovered materials

In relation to the latter point, the extent of contamination of the various recovered streams will be an important factor.

An area of obvious concern is that relating to health and safety issues, especially when records of past waste deposits are not as accurate as they could be, and where extremely hazardous materials may have been deposited. Savage and Diaz (1994) state that "based on the limited historical record and on several limited evaluations LFMR appears feasible with respect to occupational and public health and safety". Nevertheless, the risk posed to the health and safety of the workers and the general public will vary in a site-specific way according to the factors identified above, and measures taken to control emissions, and to monitor for other potential hazards such as toxic and radioactive materials.

The success of LFMR will undoubtedly be dependent upon the economic feasibility, but there will also be a need to ensure that health and safety issues can be adequately addressed.

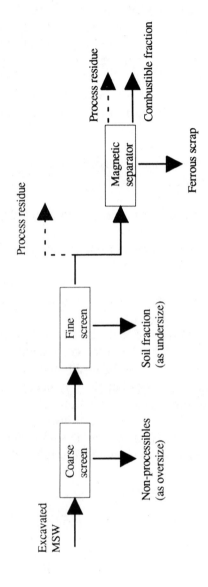

Figure 4 Landfill mining and reclamation - simple processing (Savage and Diaz, 1994)

Reproduced with Permission

CHAPTER 3

BIOLOGICAL, CHEMICAL, AND PHYSICAL PROCESSES WITHIN LANDFILL.

3.1   Introduction

The deposition of waste materials containing biodegradable matter invariably leads to the production of gas and leachate, the composition of which will vary according to the nature of the material being degraded and the surrounding environmental conditions.

The composition of landfill gas and leachate are very closely related, as each is dependent upon the nature of the emplaced waste and the activity of the micro-organisms within the landfill.   An understanding of the production, composition, and properties of each is critical to the understanding and control of the environmental problems of waste disposal to landfill, and within a risk assessment framework, forms part of the first stage (hazard identification) process. In the source-pathway-receptor model described earlier, identification of production, composition and properties of the gas and leachate help to define the hazard.   Many factors are known to affect these parameters, including waste composition, moisture content, pH, and waste particle size and density.   As it is not possible to contain waste, or its products, within landfill for ever (irrespective of the containment system specification), and because, once released, we cannot guarantee that sensitive receptors will not be affected, control of the hazard is the main area where the overall risk associated with landfill disposal of waste may be reduced, and thus ensure that this essential waste management option is as sustainable as possible.

Within waste disposed to landfill, vegetable matter, paper and cardboard and to some extent, textiles are all biodegradable.   Although, as shown in Table 6, the composition of municipal refuse varies from country to country and will vary from season to season.   In the developed world it typically contains about

60% carbohydrate, 2.5% protein and 6% lipid, the balance being comprised of "inerts" and plastics. Carbohydrates therefore comprise approximately 85% of the **biodegradable** material within municipal refuse, the overall breakdown of which can be represented by the equation:

$$C_6H_{12}O_6 \rightarrow CH_4 + CO_2 + \text{Biomass} + \text{Heat}$$

(Carbohydrate) (Methane) (Carbon dioxide)   (Bacteria)

Methane gas is a high energy fuel with approximately 90% of the energy stored in carbohydrate being retained in the methane. The conversion of carbohydrate to methane is therefore a highly energy efficient process, and much of the energy stored in the carbohydrate is contained within the methane gas. Because of the high energy value, the methane can be used beneficially as a heating fuel and for energy production.

The production of landfill gas arises as a result of bacterial activity which effects the degradation of the organic material, and the earliest reports of methanogenesis in landfill date back to the 1940's (Carpenter and Setter, 1940; Eliassen, 1942; cited in Gendebien et al, 1992). Although many of the processes thought to occur within landfill have not been proven, the presence of predicted intermediate products and end products of degradation, together with the presence of relevant enzymes leads us to conclude that the degradation of organic waste in the landfill environment is similar to the degradation of organic materials in other anaerobic environments, although the bacteria effecting degradation are likely to be different.

When deposited within the landfill, oxygen entrapped within the void spaces is rapidly depleted as a result of biological activity, and thus each of the degradation reactions occurs under anaerobic (oxygen-free) conditions, encouraging the growth of anaerobic micro-organisms, especially bacteria. Carbon dioxide and methane are produced as a result of anaerobic microbial activity, and displace nitrogen remaining from the entrapped air.

In the initial phase of degradation (fermentative/hydrolytic stage), organic material is broken down to small, soluble molecules including a variety of sugars. These soluble molecules are then broken down further to hydrogen, carbon dioxide, acetic acid and a range of other organic acids which may be subsequently converted to acetic acid (acetogenic phase). Within landfill, acetic acid, propionic acid and butyric acid have all been detected within leachate (Senior and Shibani, 1990) and this evidence together with the knowledge that acid and hydrogen-producing fermentative bacteria are also present, (ETSU, 1990) indicates that processes similar to those described in Figure 5 occur in the landfill environment.

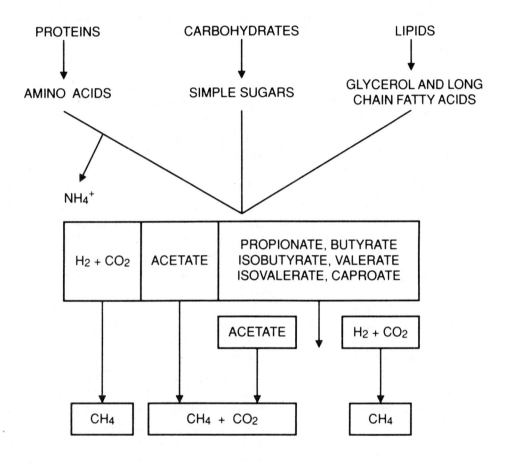

Figure 5.   Decomposition processes in domestic waste (DoE, 1992)

Reproduced with permission

Acetic acid, hydrogen and carbon dioxide form the major substrates for growth of the methanogenic bacteria (methanogenic phase). Methane gas is the end product of this final stage and together with carbon dioxide forms the major component of landfill gas. Eventually, a dynamic equilibrium is reached with a typical gas ratio within the landfill of approximately 60:40 methane: carbon dioxide.

Acid products that are not converted to gas can make up a major fraction of the leachate, and can occur at relatively high levels during the early stages of landfill waste degradation. The acidic nature of the leachate can cause the pH of the landfill to fall below a level at which methanogens are able to grow. However, the natural buffering capacity of landfill helps to stabilise the pH of landfill at a level conducive to the growth of the methanogens.

Once established, the methanogenic bacteria will help to control the pH at or near neutrality such that acetogenic phases (pH 5 - 6) and methanogenic phases (pH 6 - 8) of landfill can be clearly discerned. At this stage, landfill gas production will increase as a result of the bacterial activity, with the result that carbon compounds that might otherwise contribute to leachate TOC may be converted to landfill gas with a resultant decrease in this parameter within the MSW leachate (Table 6). The leachate hazard will therefore be reduced with a consequent increase in the landfill gas hazard.

In the later stages of landfill stabilisation, the production of landfill gas can be expected to continue until the biodegradable material is depleted or until the balance of micro-organisms is disturbed. While carbohydrate sources are present in landfill, the potential for gas remains and while gas production rates are high, it may, according to local circumstances, be economically feasible to install gas abstraction schemes for energy production. Under these circumstances, the conversion of leachate BOD to landfill gas and the abstraction of methane for energy utilisation will limit environmental problems caused by both the release of methane to the atmosphere, and escape of leachate to ground and surface waters.

The pattern of gas production assumed to be typical of landfill is shown in Figure 6 and is based on that proposed by Farquhar and Rovers (1973). In this figure, no scale is included on the "x"-axis, for the length of time taken for each process will vary considerably from site to site, according to a number of factors including, waste composition and local environmental conditions. Therefore, the timescale over which pollution events can occur as a result of gas migration will also be site dependent and could vary between a few years and more than one hundred years. For effective pollution prevention, it will be necessary to monitor and exert control for a similar time period, with important cost implications.

Effective site planning in the early stages of landfill design coupled with good design and operational procedures will help to more effectively ensure safe landfilling of waste in a cost-effective manner, with reduced long term pollution potential.

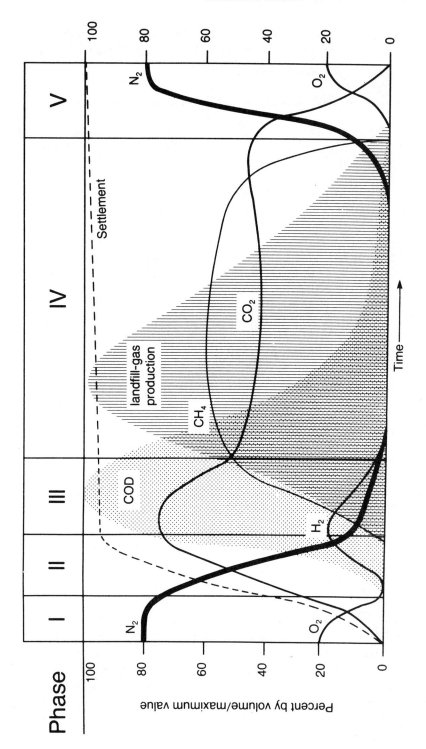

Figure 6.   Landfill stabilisation, gas and leachate production (DoE, 1993)

Developed from Pohland (1986) and Farquhar and Rovers (1973) by Wrc

Table 6.   Typical composition of leachates from domestic wastes at various stages of decomposition - all figures in mg.l^{-1}, except pH values (DoE, 1993d).

Determinand	Fresh Wastes	Aged Wastes
pH	6.2	7.5
COD	23800	1160
BOD	11900	260
TOC	8000	465
Volatile acids (as C)	5688	5
NH_3-N	790	370
NO_3-N	3	1
Ortho-P	0.73	1.4
Cl	1315	2080
Na	9601	300
Mg	252	185
K	780	590
Ca	1820	250
Mn	27	2.1
Fe	540	23
Cu	0.12	0.03
Zn	21.5	0.4
Pb	0.40	0.14

Reproduced with permission

Because of the heterogeneous nature of landfill, the bacterial population and therefore gas ratios, gas production rates and leachate composition can change from one region to the next, and can affect the landfill pollution potential. Other factors of major significance include moisture content, pH and temperature. The available surface area and the nature of the surface of refuse particles will also be important determinands of microbial activity, and will thus affect the nature of the hazard posed by landfill gas and leachate.

3.2   Factors Affecting Gas and Leachate Production

3.2.1 Waste Composition

The nature of the deposited waste material will affect gas and leachate production and composition by virtue of the relative proportions of degradable and non-degradable components, the moisture content, and the specific nature of the biodegradable element. Clearly, in a truly inert landfill site in which only non-degradable material (such as bricks and sand) have been emplaced, there will be no potential for gas production except through material carried in within infiltrating liquids, although in these circumstances, it is highly unlikely that gas will be produced.

The waste composition will affect both the bulk gases and the trace components, while the amount and composition of landfill leachate will be affected by both the absorbent and adsorbent capacity of the waste, and the waste composition. The leachate will be comprised of both the readily soluble components of the undegraded waste and the soluble products of waste degradation.

The typical composition of municipal solid waste (MSW) from the UK, an Asian city and a Middle East city is shown in Table 7, and shows that for the UK, a large proportion of the waste is biodegradable. Assuming that the 'unsorted fines' are biodegradable, then the total proportion of waste that is biodegradable is equivalent to 65.8% (this includes: paper, putrescibles, unsorted fines, garden waste, textiles and wood).

However, for many developed countries, the relative proportions of MSW components have changed considerably during the 20th century, where typically the amount of ash and cinders produced through the use of coal fires has decreased and the amount of paper and putrescible fractions have increased. In the UK, the introduction of the Clean Air Act (1956) resulted in a decrease of the amount of combustible waste burned in the home, which both decreased the ash content and increased the content of degradable material within household waste arisings.
This increase was exacerbated by the increase in the amount of office paper waste generated (McEntee and Jelley, 1991). Such changes have had a significant impact on the nature and rate of emissions from landfill sites, and clearly demonstrates the impact of changing waste composition on landfill-related hazards. Prior to 1956, most domestic waste disposed to landfill in the UK would have been ash from the fire grate; it would have a low organic content, and therefore low potential for production of landfill gas and high-BOD leachate. Subsequent to 1956 the potential for landfill gas and high-BOD leachate increased considerably.

Again using the UK as an example, many landfill gas-related problems have now been recognised, but as identified in Chapter 2, only a relatively small number of leachate-related problems have been identified. Recognising the change in potential for leachate pollution after 1956, and at the same time recognising that groundwater flow is extremely slow, it is **possible** that those few identified leachate pollution incidents are the first of many - only time will tell. Although many of the geological strata within the UK is well suited to landfill disposal and may also be capable of attenuation (in which case pollution-related incidents may be few), the potential for as yet unrecognised groundwater contamination, highlights the importance of both the need for landfill siting, design, and operation on an assessed risk basis, and the need for sustainable landfill development.

Table 7.   Comparison of waste characteristics from different countries (Holmes,1992)

	United Kingdom	Asian City	Middle East City
Vegetable	28	75	50
Paper	37	2	16
Metals	9	0.1	5
Glass	9	0.2	2
Textiles	3	3	3
Plastic	2	1	1
Miscellaneous	12	18.7	23
Weight/person/day	0.845kg	0.415kg	1.060kg
Density kg/m³	132	570	211

Reproduced with permission

The data presented in Table 7 clearly show significant differences in both waste composition and the rate of waste generation in different countries. Such changes are a function of national culture and lifestyle and will be very different in "developed" and "developing" countries. According to Holmes (1993) "Quantities of waste are invariably lower in developing countries because of lower prosperity and consumption as well as extensive scavenging by beggars and the very poor. Densities of waste are much higher because of the absence of paper, plastics, glass and packaging materials and hence a much greater concentration of putrescible matter. Moisture contents at 40-50% are much higher than those in developed countries at 20-30%". This informal sector recycling, and the effect on waste composition, shows how much different waste composition in the developed world would be if recycling activities were as efficient as those in the "Asian city" of Table 7. If it were possible to compost the vegetable matter of the Asian city (which constitutes 75% of the total waste arisings), either as a pre-treatment to landfill, or for the production of a marketable compost, then the total waste arisings for final disposal would be minimal. If the developed world could achieve similar recycling and treatment targets, then waste-associated problems would be considerably less than they are today.

3.2.2 Moisture Content

Most microorganisms including bacteria require a minimum of approximately 12% (by weight) moisture for growth, and thus the moisture content of landfilled waste will be an important factor in determining the extent of gas and leachate production. In fact, for biodegradble waste, moisture content is probably the single most important factor affecting gas production in landfill. Published data indicate that moisture levels of between 40% and 50% are optimum for anaerobic degradation (Chian and DeWalle, 1977) while EMCON (1975) and Leckie (1979) showed that adjusting refuse to field capacity accelerated decomposition. Rees has also shown (1980) that the log of the rate of gas production is directly proportional to the percentage of water content of refuse (Figure 7), and work by Klink and Ham (1982) showed that methane production increased by 25% to 50% during moisture movement through refuse compared to a situation with no movement but the same moisture content.

It is evident therefore, that landfill gas production may be increased significantly by the addition of water, but that the volume of leachate will necessarily increase. The addition of water is a key requirement of the bioreactor landfill, and arguments relating to the advisability and technology of this landfill principle are discussed in Chapter 7. If the rates of waste degradation within landfill are to be increased, then optimisation of moisture content and flow is essential. For this to be achieved, effective control and treatment of leachate will be necessary. However, the addition of excess moisture may not be necessary: The moisture content of municipal solid waste is typically 20-30% in developed countries and 40-50% in developing countries (Holmes, 1993). Upon emplacement, the moisture present within lower layers of waste will be squeezed out by the weight of waste above. The collection of the leachate and subsequent redistribution throughout the waste will encourage an even distribution of moisture, and will redistribute nutrients throughout the waste, aiding microbial activity. (Because of the heterogeneity of MSW, moisture may not be evenly distributed in the waste and in some areas can be present at levels below that required for effective microbial growth. Consequently, gas and leachate production will occur at different rates within different areas of waste and will affect the degradation processes and the long term predictability of landfill gas production).

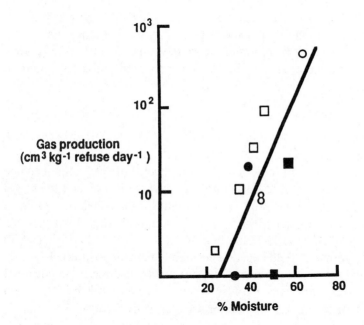

Figure 7.   The relationship between refuse moisture content and gas production (Rees, 1980)

In exceptional circumstances, excess moisture **may** also, decrease landfill gas production, especially in shallow, relatively poorly insulated sites. In this case, the high heat capacity, and high thermal conductivity can decrease the temperature and insulation properties of landfill and thereby decrease microbial activity and gas production. Without gas production, the rate of waste stabilisation will also be low and the long term pollution potential will be increased accordingly. Although in this case, hazards consequent upon gas migration may be reduced, the hazards related to the migration of leachate may become more significant. Furthermore, if too much water is present in the early stages of landfill stabilisation, then excessive leaching of soluble sugars may exacerbate the leachate pollution potential by increasing the COD and BOD loading.

Thus for rapid waste stabilisation, the level of moisture within the waste should be optimised to enhance microbial activity, thereby enhancing the rate and extent of waste degradation. Enhancement of waste stabilisation through moisture control will also enhance the rate and extent of gas production. However as indicated previously, (and as described in detail in Chapter 4) the control and management of landfill gas will be more easily undertaken than the control and management of leachate. This is especially important when considering the deleterious effects upon the environment that the escape of leachate would cause, and the cost of treatment of large volumes of leachate, especially after dilution within underground water bodies.

When viewed as a biological reactor, the landfill can be equated with a stationary fixed-film reactor in which poor mixing is the most significant constraint on bacterial activity, and hence waste degradation. Physical mixing of the waste within the site is clearly impractical, and thus leachate recycle provides the only means of addressing this issue. Also, because moisture movement has been shown to increase the rate of gas production within landfill, leachate recirculation has been proposed as a means of enhancing the rate of waste degradation/stabilisation within landfill, and which will at the same time, treat the leachate as it passes through the waste. Leachate movement could be anticipated to increase waste degradation rates as a result of better moisture distribution throughout the waste (creating more areas within the waste where bacteria may grow), enhanced access of bacteria to soluble nutrients, and removal of inhibitory compounds.

Working with shredded refuse, Pohland (1989) demonstrated that the readily accessible and reactive waste components are more effectively and predictably converted and removed under the influence of leachate recycle than with single-pass leaching, while Buivid et al (1981) showed leachate recycle to be most effective after addition of anaerobic sludge and calcium carbonate. Barlaz *et al* (1987 and 1989) concluded that leachate recycle without neutralisation stimulated the formation of acid products, but that neutralisation of the leachate decreased acid production and enhanced gas production. These laboratory studies used shredded waste which has been shown to increase acid production of its own accord and which is atypical of most operational and closed landfills, where a larger waste particle size, scale factors, and altered buffering capacity may limit potential for acid production. Larger scale studies at the Brogborough test cells, containing approximately 15,000 tons waste (Knox, 1995), showed that the addition of water alone significantly enhanced both the rate of gas production and the concentration of methane. However, even in studies of full-scale landfills the effectiveness of leachate recirculation remains uncertain (Doedens and Cord-Landwehr, 1989).

Despite these rather ambivalent data, leachate recycle appears to offer many advantages to landfill management (as currently practiced) and towards the attainment of more sustainable landfill practices. As in many other areas of landfill research, early studies were conducted on a fairly small-scale, and it is only in the last ten years that work has been conducted on a scale similar to that of a working landfill. As well as problems associated with scale differences, the extreme heterogeneity of landfilled waste creates problems when working at small-scale, where the attainment of a small but representative sample is essentially impossible to achieve. Results from larger scale studies which are currently in progress, such as the Brogborough test cells (Campbell and Croft, 1991), are likely to produce more meaningful data.

3.2.3   Waste Particle Size and Density

Within landfill management, the major means of controlling particle size is through shredding: the arguments for shredding being that it increases homogeneity, increases the surface area/volume ratio (to aid microbial degradation), and reduces the potential for preferential liquid flow paths through the waste. However, most studies that have investigated the effects of shredding (and thus particle size) have shown a negative affect on waste degradation processes (EMCON Associates, 1975; Chian *et al*, 1977; Buivid, 1980). The reason for the above appears to be an enhancement of the acid-producing phase, resulting in the production of a low pH, high COD leachate that inhibits methanogenic processes. Pre-composting of waste to remove to remove the easily biodegradable components may reduce the potential for acid production and offer an alternative landfill management technique for enhanced waste stabilisation. However, although composting has been shown to enhance waste degradation in some cases, the reasons for this are not clear.

The absorption of water by refuse will be affected by particle size and density and is a function of the waste composition and landfill operating procedures. Using waste with a high moisture content, Rees and Grainger (1982) showed that increasing the waste density from 0.2 to 0.47 tonnes/m^3 decreased gas production: this was ascribed to stimulation of acid production. In similar studies with drier waste, Rees and Grainger (1982) showed that increasing the waste density from 0.32 to 0.47 tonnes/m^3 significantly increased gas production. There are many factors that could influence these results. For example, it seems feasible that greater compaction could be achieved in high moisture-content waste, that would reduce the surface area/volume ratio for microbial attack. Alternatively, the extent of water absorption per unit mass is likely to increase at high surface area to volume ratios (low packing densities); and at low packing densities, the increased moisture content and surface area of the refuse could be expected to favour microbial growth and gas production.

Particle size will also influence waste packing densities, and particle size reduction (by shredding) could increase biogas production through the increased surface area available to degradation by bacteria. However, the creation of smaller particles allows the establishment of higher packing densities (although moisture content of the waste will also affect this) which may decrease water movement, bacterial movement and thus bacterial access to its substrate. Clearly, much of the above is speculation, and despite many years of study of landfill microbiology, the relationship between particle size and moisture within full-scale landfills is not well characterised, but it is evident that whatever the cause, refuse density is one of the landfill characteristics which directly affects gas production, and which will therefore affect landfill pollution potential. Optimisation of particle size and packing density could be anticipated to increase gas production and the associated (more easily controlled) gas hazard and therefore minimise leachate 'strength' and the associated hazard caused by leachate migration. However, we are as far away from knowing these optima now as we were twenty years ago. One of the main reasons for this is that for landfill operators who wish to maximise profits, maximum waste densities rather than optimum waste densities are the goal. However, any short term gains may be made at the expense of long-term losses associated with longer post-closure stabilisation periods, and the potential long-term monitoring and remediation costs.

3.2.4 Temperature

Microbial activity is affected by temperature to the extent that we are able to segregate bacteria according to their optimum temperature operating conditions. Thus factors affecting landfill temperatures are likely to affect microbial activity and hence gas production. An increase in temperature tends to increase gas production (Massman *et al*, 1981; Pacey, 1986; Emberton, 1986)

Landfill temperature is initially influenced by the exothermic (heat-releasing) microbial activities during the initial aerobic processes in landfill stabilisation and the initial waste temperature has been directly related to the refuse temperature at placement (Farquhar and Rovers, 1973). Temperatures as high as 60 - 80^{0}C may be attained during this phase (Ehrig, 1981; Pacey and Karpinski, 1980). It has also been shown by Pfeffer (1974) in laboratory studies with domestic refuse, that degradation was much greater at thermophilic (60°C) rather than mesophilic (42°C) temperatures. However most landfills operate in the range 15 - 45°C and optimum temperature ranges of 30 - 40°C have been reported (Pacey, 1986).

Waste temperature control for optimum gas production has not been conducted on a large scale to date, but may be considered as one option for increasing gas production while decreasing the leachate hazard through reduction of the leachate carbon content. The potential for temperature control will be greater in contained landfill sites and may be exercised in a number of ways including air injection to encourage temporary composting conditions (although fire hazards may be associated with this method), and recirculation of leachate after heating. Large-scale studies of the effects of air injection are currently in progress at the Brogborough test cells in the UK, and have shown both an increase in the rate of gas production, and an increase in the methane content of the landfill gas (Knox, 1995).

3.2.5 pH

The activity of all microorganisms is affected by pH, and the methanogenic bacteria within landfill will only grow within a narrow pH range around neutrality (Fielding *et al*, 1988). Results from recent studies (ETSU, 1990) indicate that the cellulolytic bacteria isolated from landfill have broader pH ranges than the methanogens. Most bacteria isolated in these studies were able to grow between pH 5.5 and 8.0 although the optimum pH was between 6.8 and 7.7. Thus although disturbance of pH can potentially affect landfill gas production, it is likely to exert a greater effect on the methanogenic bacteria than the cellulolytic bacteria.

Acidic conditions within landfill result mainly from the generation of organic acids, which have been shown to contribute as much as 95% of the TOC during the early acid-producing stages of waste degradation (Harmsen,1983), and the close relationship between pH and microbial activity/landfill stabilisation is such that pH has been found to be the best predictor of methane generation rates (Segal, 1987). In terms of leachate composition, landfill development is often described as being in the acetogenic leachate phase or the methanogenic leachate phase. The stage of development is determined simply by measuring the pH, and can provide useful information for overall landfill mnanagement. The switch from acetogenic to methanogenic phase can occur very quickly, a factor which is difficult to explain, and which if understood, could facilitate more effective control of waste degradation processes generally.

The information supplied above shows that waste composition, moisture content, waste particle size, waste density, temperature and pH will all affect both the extent and the rate of waste degradation.. However, very little of this knowledge has been utilised by landfill operators in their management of landfill sites: until control of these parameters is attempted on an operational landfill, it will be impossible to know to what extent the bioreactor landfill can meet the requirenments of a sustainable landfill, and reduce the risks of harm and pollution to an acceptable level. The problems of transferring information from small-scale laboratory experiments to a full-scale landfill has been briefly mentioned above: the effect of any of these parameters may be significantly altered on an operational landfill. However, for a landfill operator, it is difficult to justify investment in plant and in new management and operational procedures when these uncertainties exist. At the same time, the view of some Govenments that this research should be funded by the landfill industry itself, is also understandable. The cost of the necessary studies would not be insignificant, and it seems that as the problem is common to many countries of the world, an international collaborative programme, with well defined objectives would be an appropriate way forward. In the absence of such work, the pre-treatment of waste prior to landfill may become the only allowable means of managing the long-term risks of landfill.

A number of reviews of the parameters affecting gas generation have been conducted and include those by Senior and Kasali (1990), and Barlaz *et al* (1990).

3.3   Physico-chemical Processes in Landfilled Waste

The leaching of toxic species from a deposit of waste will also be affected by waste-related and environmental physico-chemical processes. Some layers of waste may be hydrophobic or impenetrable to leachate and the material may remain *in situ* in a relatively undegraded state for many years. However, once toxic species are in solution, reaction with the environment can occur in a number of ways.

The main physico-chemical processes that occur in landfill are presumed to be precipitation, adsorption, ion-exchange and volatilisation, although very few studies of the physico-chemical processes in landfilled waste have been undertaken. Volatilisation will affect landfill gas composition, but otherwise these processes will have a major impact on the leachate composition and the associated hazard. Leachate composition and reactions will also be affected by the waste characteristics including the redox potential, adsorption characteristics, temperature and biological mechanisms.

3.3.1   Precipitation

Precipitation is an important factor in leachate attenuation processes within landfills and is especially important in attenuating heavy metals. Most heavy metals are characteristically more soluble at low pH values and therefore may be expected to be more mobile during the early stages of waste stabilisation. However, as the waste decomposition proceeds, the pH rises causing the solubility of most heavy metals to decrease. Also under conditions of lower redox potential, the precipitation of heavy metals as insoluble sulphides and carbonates may occur (Rees, 1982), thus reducing the concentration of heavy metals in the leachate. According to Pohland (1991) inorganic heavy metals codisposed with MSW are attenuated by the microbially-mediated processes of reduction, precipitation, sorption and waste matrix capture and in general heavy metals do not pose a great problem either at domestic sites or at co-disposal sites. Harmsen (1983) has shown a large reduction in leachate heavy metal concentration after the onset of methanogenesis compared to the concentrations during the acid-producing phase of waste degradation, thus providing evidence to support the theories above. In some ways, the precipitation of metals in this way could be considered as returning these compounds to the earth in a form which is similar to that in which they were originally present before mining, and may thus represent their most appropriate and sustainable means of disposal.

However, current knowledge of the long-term stability of the landfill environment is uncertain and thus the long-term stability of precipitated metals is also uncertain. The situation may be further complicated by the action of complexing agents such as ammonia, and organic compounds containing nitrogen, oxygen and sulphur, which can form soluble complexes with metals and thus promote leaching. Thus the humic and fulvic acid products of waste decomposition could facilitate the leaching of trace metals from landfill. Despite this, Belevi and Baccini (1989) showed that more than 99.9% of metals remain in landfill at the end of the intensive reactor period.

Precipitation reactions will therefore clearly have an impact on the potential pollution arising from the migration of leachate from landfill.

3.3.2 Adsorption
Adsorption is the process whereby a chemical species moves from one phase and is held on the surface of another where it accumulates. Chemical adsorption, or chemisorption, involves molecular interaction, while physical adsorption relies on bonding due to electrostatic forces (e.g. van der Waals forces). Adsorption may be important in the removal of ammonia from the liquid phase under anaerobic conditions and may therefore also be important in the immobilisation or remobilisation of metals.

3.3.3 Ion-exchange
The strength of the electrostatic bond between adsorbate and adsorbent varies according to factors such as pH, temperature, redox potential, salt concentration, and charge density. Adsorbed ions will be displaced and readsorbed according to the local conditions and their movement will thus be slowed or stopped. Some ionic species such as chloride tend not to undergo ion exchange and in this case their movement through the landfill will be dependent upon other factors. The unreactive nature of the chloride ion means that it is often used as a marker of the front of a landfill leachate pollution plume.

In the landfill environment, ion-exchange of heavy metal ions may occur, although these processes are likely to be limited by the competing effects of alkali metals and alkali earth metals such as Na, K and Ca, and ion exchange is not likely to be a major attenuating factor in this environment. The ion exchange of heavy metals may have a greater impact in natural strata after escape of leachate from a landfill, and has been used in justification of landfill construction on an attenuate and disperse principle (Philpott *et al*, 1992). In this case, boulder clays were judged to have significant cation-exchange capacity, and the site in question was designed such that leachate could seep through the 5m basal blanket at such a rate that it would be attenuated to an acceptable quality. Contingency measures for leachate collection, treatment and disposal were also installed.

3.3.4   Volatilisation

Compounds such as volatile organic compounds will partition between the vapour and liquid phases within the landfill environment, according to their vapour pressure. Those compounds within the gaseous phase may then be removed from the landfill through surface emissions or via gas abstraction systems, and will at any given time affect the composition and associated hazard of migrating landfill gas, and the hazards due to surface emissions. In the USA, statutory measures introduced in 1991 under the Clean Air Act have been designed to control fugitive volatile emissions from landfills; the target pollutants are non-methanogenic organic compounds (NMOC's) and methane on MSW landfills with NMOC emission rates in excess of 150 tonnes per year. No such controls are currently in place in Europe.

A combination of the above processes together with biological degradation processes described previously, will modify the composition and associated hazards of both landfill gas and leachate. An understanding of these processes and their potential effects are essential to the design and operation of a sustainable landfill, and reduction of the long-term landfill-associated environmental risks. The relative importance and effectiveness of each of the various factors described above will vary according to the nature of the emplaced waste, and will be an important factor when choosing between the pre-treatment landfill option, or the bioreactor landfill for sustainable landfill development (discussed further in Chapter 7).

Optimisation of the parameters that enhance waste degradation (and stabilisation) processes, will result in enhanced microbial activity within the waste, and more active methanogenesis. Under these conditions, the co-disposal of industrial and hazardous waste will be more effectively achieved.

3.4 Co-disposal

Co-disposal (referred to in the European Landfill Directive (CEC, 1991) as joint disposal) is "the disposal, in landfills, of a restricted range of industrial waste (including some special waste) together with decomposing municipal waste or similar degradable waste, in such a way that the industrial waste gradually undergo a form of treatment." (DoE, 1994a). The underlying concept of co-disposal is that waste are **treated** as a result of the microbiological and physico-chemical activity within the landfill. This treatment reduces the hazard associated with the waste, and hence the long-term risks. In this way, it could be argued that **successful** co-disposal represents a sustainable waste treatment operation, that upon ultimate liner failure, allows the reassimilation of the waste into the surrounding environment, at an acceptable risk.

The establishment of stable methanogenesis within a landfill creates an environment in which the treatment of waste which might otherwise create disposal problems, can occur. In these circumstances, the codisposed waste may be degraded by the bacteria present within the waste to relatively non-toxic products. For example, on the basis of a number of laboratory experiments Pohland (1991) concluded that "Inorganic heavy metals codisposed with MSW are attenuated by the *in situ* microbially-mediated physico-chemical processes of reduction, precipitation, sorption and waste matrix capture. Similarly, codisposed organic waste constituents will fractionate according to their physical and chemical properties and those of the waste matrix, and are retained or transformed primarily by sorption and/or bioconversion..." The treatment capacity of refuse is therefore dependent upon physico-chemical and microbiological processes, but in many cases the actual mechanisms of waste treatment are not known. The treatment process may degrade the waste to relatively non-toxic products, it may neutralise waste, or may render the waste immobile.

With the exception of the disposal of sewage sludge, and non-hazardous MSW combustion ash, co-disposal is not allowed in the USA (where all hazardous waste has to be pre-treated), and according to the proposed Landfill Directive (CEC, 1991), will be banned from all European sites, except those with existing licences. Thus although co-disposal looks fine on paper, there is a strong anti-co-disposal lobby that will mean that its continued use will be uncertain. One of the major problems facing the continued use of co-disposal is the lack of evidence to show that the theory works in practice, for it is only in relatively recent years that data of the quantity and quality needed has been available. Failure to control co-disposal could lead to obvious environmental harm, but as identified above, the types and nature of the attenuating processes are recognised only in general terms. As a result, fine tuning of the co-disposal process is not possible and effective management is therefore essential. For many years the management of co-disposal practices has been weak, yet there have been few reports of problems associated with co-disposal. It is possible that its detractors see it as a cheap process (and therefore it can't possibly be as good as higher cost processes!) yet if undertaken in a properly controlled manner, co-disposal is not a cheap option. Even if it were, effective processing of waste at relatively low costs would certainly facilitate the "development" within "sustainable development".

Many of the wastes that are codisposed are toxic if left untreated, and the control of their co-disposal should be exercised through clear identification of the nature, amount and concentration of incoming waste and subsequent landfilling within accepted loading limits by safe and effective means.

Within the UK, a draft document for consultation has been produced (DoE, 1994a), which specifies requirements for effective co-disposal, including site requirements, waste acceptability criteria, loading rate considerations and other relevant factors. This document (WMP 26F Landfill co-disposal: draft for external consultation), after the consultation process and subsequent amendments will form the basis of UK guidance on co-disposal. It indicates that for some wastes (primarily inorganic solid waste and some aqueous waste containing acids, heavy metals, degradable organics, and low concentrations of other readily attenuated components), co-disposal represents the BPEO. There are many arguments for and against co-disposal, but it does seem that any process that allows reassimilation of a waste stream into the environment at an acceptable risk, should be encouraged. Using one of the above waste streams as an example, Figure 8 shows possible materials/products that would be produced if disposed via various means, including co-disposal.

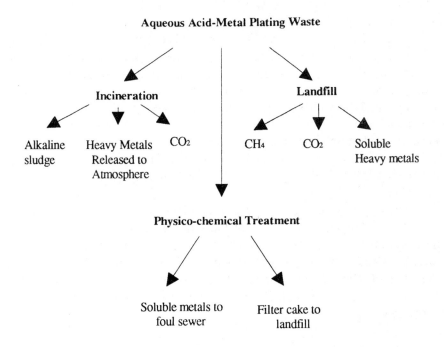

Figure 8. Some emissions to the environment form the disposal of Aqueous acid metal-plating waste

Dilute acid waste seldom occurs in uncontaminated form, and in the example shown has been assumed to contain soluble heavy metals e.g. from a metal plating works. Incineration of such waste would be extremely inefficient, as much of the energy would be used to effectively convert water to steam. The alkaline sludge produced during the gas clean-up would require disposal, while heavy metals and carbon dioxide (amongst other gases) would be released to atmosphere. Physico-chemical treatment would react the acid waste with an alkaline waste to produce a liquid waste for discharge to sewer (with concomitant environmental consequences), and a sludge for filtering/drying, and disposal of the filter cake to landfill. Energy would be required to effect the mixing and separation stages of treatment. However, co-disposal to landfill would require much less energy than the two previous processes, and would produce a relatively insoluble metal precipitate within the reduced environment of the landfill, plus carbon dioxide and methane (as well as other gases that would vary according to the specific type of acid). Although gas abstraction at landfills is relatively inefficient, the methane produced could be used for energy production. This rather oversimplified example demonstrates that consideration of BPEO is not a simple exercise (even for a relatively simple waste stream), but that co-disposal does appear to offer a number of advantages over other options.

If co-disposal is not carefully undertaken then the risk associated with leachate migration from the landfill can be considerably greater than that for "ordinary" landfill sites. However when co-disposal is undertaken effectively, compounds present in "co-disposal" landfill leachate are often little different from those generated from MSW alone (e.g. DoE, 1994). Some concern has been expressed over the long-term solubility of, for example, immobile metals when the landfill becomes aerobic at some time in the future. Yet the landfill environment is unlikely to become any more aerobic than any other terrestrial environments at depth (which is relatively low), and overt concern over this issue would seem to be misplaced.

In a review of the co-disposal of phenols, cyanides, acids and heavy metals, Knox (1990a) concluded that the removal capacities for all four groups were favourable when compared to other types of anaerobic reactor and that there is clear potential to treat significant volumes of waste containing these chemicals and a range of other organic compounds. The conditions under which degradation occurred optimally were similar for all four groups of chemicals and were as follows (Knox, 1990a):

- saturated conditions
- high irrigation rate/fluid velocity
- established methanogenesis
- elevated temperatures
- long hydraulic retention time (HRT)

These conditions are the same as those that may be anticipated to promote maximum rates of waste stabilisation and to therefore reduce the long term pollution potential of landfilled waste. The use of highly engineered and specified containment landfills will create the opportunity to optimise (within financial constraints) the limits of the above conditions to more effectively and safely codispose different waste streams, especially as Knox (1990a) was able to demonstrate the robustness of the methanogenic waste degradation process under adverse conditions of co-disposal.

For effective co-disposal, the draft document for consultation (DoE, 1994a) states that "flushing bioreactor conditions are needed .... This leads to a separate requirement for containment and possibly recirculation" (see Chapter 7). Under enhanced conditions of gas generation (and hence microbial activity) the potential for effective co-disposal will be enhanced accordingly. Although at the moment very few landfill sites undertake leachate recirculation, it is a concept that is receiving a great deal of attention and one that appears to facilitate both co-disposal practice and more sustainable landfill development.

One of the perceived problems of co-disposal is the accumulation within leachate of undegraded codisposed waste. This may occur when the codisposed waste is added at a loading rate above that which the bacterial populations can tolerate, with the result that bacterial growth and activity is inhibited or stopped and the codisposed waste passes through the site untreated. Strict adherence to loading rate guidelines is therefore an important feature of co-disposal, for not only may the co-disposal waste remain untreated but the normal microbial landfill processes may also be inhibited. If methanogenesis were to fail then the landfill could become acidic, which may remobilise heavy metals within the waste, and thereby add to the problem associated with the leachate. Upon failure of the containment system only the natural attenuation mechanisms of the surrounding strata would help to alleviate the polluting effects of the leachate.

The practice of co-disposal is currently being debated within the European Community where there are conflicting views on the merits of co-disposal as a waste treatment option, but as identified above it will only be allowed to continue at those sites currently licensed to do so; No new co-disposal licenses will be issued. In those countries whose landfill operations are better controlled, the impact of the Directive will still be significant; For example, in the United Kingdom, there are approximately 4000 licensed landfill sites where approximately 85% of controlled wastes are disposed. Approximately 2.8M tonnes of "difficult" wastes are also generated which would require alternative disposal routes if co-disposal was banned. The UK DoE has estimated that of this 2.8M tonnes, small quantities could be incinerated, and approximately 15% would be suitable for discharge to sewer after pre-treatment. This leaves approximately 2.2M tonnes for disposal to mono-disposal sites. The proposed ban on co-disposal will, according to the Department of Environment (cited in Anon., 1994b), add an extra £160 M per year to UK industries' waste disposal costs. As a result of engineering and other requirements of the Directive, the cost of landfill disposal can also be anticipated to increase significantly. As the cost increases, and the differential between landfill and other disposal options, such as incineration decreases, so the easier it becomes to use alternative disposal routes that are more favourably placed in the waste treatment hierarchy. In this way, the objectives of the European 5th Environment Action Programme begin to become achieved, but if this is achieved at the expense of effective but cheaper options, then one must question the advisability of a strict waste management hierarchy.

Also, although cost is obviously an important factor, in countries such as the Netherlands, Denmark and Japan, where local geology cannot easily support landfill, there is also a greater political will to find alternatives to landfill. However, the inability of some countries to practice co-disposal should not preclude the pratice in those that can. At the time of writing, the future of this proposed Directive remains uncertain, and may not be introduced. Under these circumstances utilisation of co-disposal would be at the discretion of the individual nations. Although knowledge of co-disposal processes is fairly superficial, this should only serve to encourage further study to optimise conditions for effective co-disposal. It will not be necessary to understand every single process - no-one would claim, for example, to be able to identify all reactions occurring in a sewage sludge digester accepting a varying sludge input, yet no-one questions the effectiveness of the process.

A list of waste streams that have been successfully codisposed, and the role of research in co-disposal can be found in Knox (1992).

CHAPTER 4

LANDFILL GAS

4.1 Production

The amount of gas produced during the degradation of landfilled waste will vary according to local conditions, but the theoretical yield can be predicted from the Buswell equation (Buswell and Hatfield, 1939) as follows:

$$C_nH_aO_b + [n-(a/4+b/2)]H_2O \rightarrow (n/2-a/8+b/4)CO_2 + (n/2+a/8-b/4)CH_4$$

Application of this equation relies upon elemental analysis of a waste and assumes 100% degradability of the waste. EMCON (1981) determined the elemental composition of MSW to be:

C	H	O	N
99	149	59	1

Using these figures, and assuming that the organic content of MSW is approximately 60% (w/w), that this contains about 10% lignin, and therefore that it should be about 55% degradable (Chandler et al, 1980; Op den Kamp et al, 1988; cited in ETSU, 1992), then the expected gas yield would be approximately 320m^3/tonne (ETSU, 1992). Work conducted by Biostrategy Associates Ltd on behalf of ETSU (1992) provides a summary of the potential methods for gas yield calculation.

However, in practice most gas abstraction schemes are based on an actual recovered yield of 100-120m^3/tonne that accounts for poor abstraction efficiency. Nevertheless, it is clear from the above that in modern landfill sites, there is

Nevertheless, it is clear from the above that in modern landfill sites, there is potential for production of large volumes of landfill gas. The nature of the hazard caused by this gas, and the potential benefits that may be derived, will be dependent upon site-specific and environmental conditions, and upon the specific gas composition.

4.2 Composition and properties

It is evident from Figure 6 that once waste has become methanogenic, the major gases present in landfill gas are methane and carbon dioxide which typically make up 50-60% (v/v) and 30-40% (v/v) landfill gas respectively. Methane is a colourless, odourless, flammable gas, with a density lighter than air. Carbon dioxide is also colourless and odourless, but is non-flammable, and is denser than air. Other "bulk" components of landfill gas (those gases measured at percentage levels) can include hydrogen, oxygen and nitrogen. When present, these latter two gases are normally indicative of air ingress within the landfill. Nitrogen is essentially inert and will have little affect except to modify the explosive range for methane. However, as the flammability of methane is dependent upon the presence of oxygen, it is important to control oxygen levels. This is especially true at landfill sites where gas abstraction and collection are undertaken, where in many cases the level of oxygen within a gas collection system is used to monitor abstraction rates, and to control the rate of gas pumping.

Typical values of the major components of landfill gas are shown in Table 8. The relative proportion of these gases, together with key trace components, often occur within a range that is characteristic of landfill gas and can be used to differentiate landfill gas from gas from different sources. The typical composition of a range of gases can be found in DoE (1986) and Williams and Aitkenhead (1991).

Explosion and Fire. Both methane and hydrogen are flammable in the presence of oxygen and are therefore potentially explosive if ignition occurs within a confined environment. Methane is flammable in air within the range 5-15% by volume, while hydrogen is flammable within the range 4.1-75%. However, except in the early stages of waste stabilisation when the concentration can reach levels of approximately 20% (v/v), hydrogen is seldom present within landfill gas at levels within the explosive range. The exact values for the upper explosive limit (U.E.L) and lower explosive limit (L.E.L) for methane may vary according the concentration of other bulk components such as carbon dioxide and nitrogen (Hoather and Wright, 1988), but for most purposes, the flammable range 5-15% (v/v) is recognised, and is the basis upon which methane gas control levels have been set. Although methane is often present within waste at concentrations above this range, subsequent dilution as a result of migration, will cause a reduction of

Table 8.   Typical landfill gas composition (DoE, 1992)

Component (% Volume)	Typical value (% volume)	Observed Maximum (% volume)
Methane	63.8[1]	88.0[2]
Carbon Dioxide	33.6[1]	89.3[1]
Oxygen	0.16[1]	20.9[1,3]
Nitrogen	2.4[1]	87.0[2,3]
Hydrogen	0.05[4]	21.1[1]
Carbon Monoxide	0.001[4]	0.09[2]
Ethane	0.005[4]	0.0139[2]
Ethene	0.018[4]	-
Acetaldehyde	0.005[4]	-
Propane	0.002[4]	0.0171[2]
Butanes	0.003[4]	0.023[1]
Helium	0.00005[4]	-
Higher Alkanes	<0.05[2]	0.07[1]
Unsaturated Hydrocarbons	0.009[1]	0.048[1]
Halogenated Compounds	0.00002[1]	0.032[1]
Hydrogen Sulphide	0.00002[1]	35.0[1]
Organosulphur Compounds	0.00001[1]	0.028[1]
Alcohols	0.00001[1]	0.127[1]
Others	0.00005[1]	0.023[1]

Notes:
1.   Data taken from Waste Management Paper No 26.
2.   Published data supplied by Aspinwall & Company.
3.   Entirely derived from the atmosphere.
4.   Taken from: Guilani, A J "Application of conventional oil and gas drilling techniques to the production of gas from garbage". American Gas. Association Transmission Conference, Salt Lake City, Utah, 5-7 May 1980.
5.   Landfill gas is usually saturated with water vapour, up to 4% by weight, depending on the gas temperature. At 25°C a value of 1.8% by weight is typical.
6.   When undertaking initial confirmatory analysis by gas chromatography, the first five compounds listed above are usually identified when looking for the presence of landfill gas.

the "within waste" concentrations to levels that will fall within the flammable range. In 1982, a UK survey identified 62 landfill-associated incidents including fires, explosions and gas migration/accumulation (Anon., 1982). Although the majority of these occurrences were associated with gas migration without serious incident, examples of gas-related explosions such as that at Loscoe (Williams and Aitkenhead, 1991) are well known. In this incident, a bungalow was destroyed as a result of the migration of landfill gas from an adjacent landfill, which collected in the cellar, and exploded when the boiler of the central heating system ignited. Gendebien *et al* (1992) cite 55 examples of landfill gas-associated incidents involving explosions, fires, or human injury from around the world, including the USA, Canada and Japan. In some of these incidents human death and injury occurred and serve to emphasise the potential risks associated with landfill gas. One of the most noted examples of a landfill gas explosion resulting in human casualties occurred in Winston-Salem, USA, where 25 guardsmen were injured. Of these, twelve suffered serious injury and three died.

Where the escape of landfill gas is not confined, the explosion risk will be lower, but there will be an associated risk of fire. As the management of landfill sites improves, the potential for fire decreases, but in the past many landfill operators accepted that minor fires were an inevitable part of landfilling waste. When fires occur within the waste itself, they can be extremely difficult to extinguish and can lead to uncontrolled and unpredictable subsidence, and the production of smoke and toxic fumes. In many cases the only way of extinguishing such fires is to remove the waste, and dig out the flammable material to the surface where it can be dealt with more easily. This process, in itself, is extremely hazardous.

Some fires may occur, in the absence of a source of ignition, as a result of spontaneous ignition. The theory has been summarised by Gendebien *et al* (1992), and requires that the auto-ignition temperature of methane (235°C) is exceeded. This can be facilitated by heating caused by decomposition processes within the waste, together with the presence of pyrophoric compounds such as heavy metals and sulphur, although clearly there are unlikely to be many circumstances within landfill where temperatures as high as this may be achieved. An excess supply of oxygen is also required and thus in well controlled landfills, fires of this nature should not occur. A number of examples of fires within and beyond landfill site boundaries have been cited by Gendebien *et al* (1992).

Asphyxiation. In this context, asphyxiation includes asphyxiation of humans, other animals and plants. The asphyxiation properties of landfill gas are related more to the ability to displace oxygen from an environment, rather than the presence of particular gas species. For humans, oxygen concentrations below 10% (v/v) may result in permanent brain damage and oxygen concentrations should not be allowed to fall below 18% (v/v) at normal atmospheric pressure (Hoather and Wright, 1988).

For plant life, the lack of oxygen rather than the toxic effects of landfill gas appear to be the primary reason for vegetation 'die-back', although carbon dioxide is toxic to plant roots at elevated concentrations, where plant sensitivity is species-dependent. Methane is non-toxic, but through displacement of oxygen within the root-zone, or as a result of bacterial methane oxidation to carbon dioxide (in which oxygen is utilised), may cause death of surface vegetation. Evidence in support of the above is summarised in Gendebien et al (1992).

Where landfill gas migration has occurred, the migration pathway can often be followed by visual observation of the surface vegetation, including trees, which show withering at leaf margins, defoliation, and branch dieback (Hoather and Wright, 1989). In extreme cases, surface heating of the soils can also be detected. Whether these surface-heating effects are a result of heat transference from the warm gas or a consequence of biological methane oxidation is not clear. In the latter case, the methane oxidising bacteria utilise the methane for bacterial growth, releasing carbon dioxide as an end product (Williams and Hitchman, 1989), and some consideration has been given to the use of such bacteria in landfill gas control systems (Harries, 1988).

The metabolic effects of asphyxiation on vegetation have been summarised by Gendebien et al (1992) and include mineral deficiency, increases in heavy metal mobility (with associated toxic effects), and the stimulation of production of toxic compounds. Environmental conditions as well as species-specific variation in sensitivity to oxygen deprivation affects the extent of damage to vegetation. There are many examples of vegetation die-back caused by migrating landfill gas. Gendebien *et al* (1992) list 31 examples from around the world, which are representative of this common problem.

Toxicity. Toxic elements of landfill gas include carbon dioxide, and numerous trace gases, to which both animals and plants may be sensitive.

Carbon dioxide is toxic, can be an asphyxiant by virtue of oxygen displacement (described above), and can cause death due to paralysis of the respiratory centres (Hoather and Wright, 1989). The threshold limit value for CO_2 is 0.5% and concentrations above 5% result in laboured breathing, headaches and visual disturbances. The long term occupational exposure limit (OEL) is 5,000 vpm and the short term occupational exposure limit (STEL) is 1.5% by volume. In most situations arising from landfill gas-associated problems, carbon dioxide toxicity will only occur when collection in an enclosed environment occurs. Under these conditions there would also be potential for explosion due to potentially high methane levels and gas control measures would be required to alleviate such a potentially dangerous situation.

Carbon dioxide can also be toxic to plants and can result in plant death. According to Rettenberger (1985, cited in Gendebien et al, 1992), concentrations above 20% are phytotoxic. The phytotoxic effects of carbon dioxide have been reviewed by Gendebien *et al* (1992).

Knowledge of the toxic effects of the trace components of landfill gas is still relatively poor and indeed, relatively few studies have been conducted on the number and type of trace gases that may be present.

Hydrogen sulphide. Hydrogen sulphide (H_2S) has a distinctive odour (normally described as "bad eggs"), and is explosive in the range 4.4-45% by volume in air. It is normally present at low levels within landfill gas but can reach concentrations as high as 35% (v/v) (DoE, 1986). It is normally associated with the degradation of high sulphate waste such as plasterboard and other gypsum-containing materials, although H_2S-related problems have also occurred at sites situated in coastal regions as a result of sea water ingress. In most circumstances, it does not normally present a hazard but it is toxic at low concentrations with an Occupational Exposure Standard (OES) of 10 ppm (8 hour time weighted average reference period), and 15 ppm Short Time Exposure Limit (STEL) (10 minute reference period). The hazard posed by hydrogen sulphide will be considerably reduced in most cases as a result of atmospheric dilution. However, death from H_2S poisoning in enclosed environments has occurred and care should be exercised to ensure the health and safety of all workers. In this case, procedures for entry into confined spaces should apply.

Trace Components The trace components of landfill gas mainly comprise a range of alkanes and alkenes, and their oxidation products (aldehydes, ketones, alcohols and esters). Waste Management Paper 26 (DoE, 1986) lists 108 compounds, or groups of compounds found in landfill gas sampled at six different landfill sites. Many of these trace compounds in landfill gas are recognised toxicants when present in air at concentrations which exceed established toxicity threshold limit values (TLVs) or the Occupational Exposure Standards (OESs) set by the Health and Safety Executive. Anyone coming into contact with landfill gas is therefore potentially at risk from the toxic nature of the minor components, and under the Control of Substances Hazardous to Health Regulations (COSHH, 1988), landfill operators are legally responsible for the health of employees and are required to comply with OES's and exposure limits set by the Health and Safety Executive (HSE).

Data presented in Table 9 compare the Occupational Exposure Level (OEL) for a number of compounds detected in landfill gas and the maximum observed concentration for each. For each compound, the maximum observed concentration exceeds the OEL, in some cases by at least one order of magnitude.

Table 9.   Compounds detected in landfill gas from UK sites which exceed OEL
value in EH40 (Clay and Norman, 1988)

Compound	OEL (mg/m3)	Maximum observed (mg/m3)
Hexanes	360	628
Benzene	30	114
Toluene	125	460
Xylenes	125	470
Propyl Benzenes	1	292
Dichloroforomethane	40	93
Vinyl Chloride	3	32
1,2-Dichloroethylenes	40	302
Tetrachloroethylene	335	350
Camphor/Fenchone	12	13

Reproduced with Permission

However, in most workplace situations, significant dilution could be anticipated as a result of mixing with air such that the maximum observed concentrations shown in Table 8, and as measured in undiluted gas, could be anticipated to fall below the relevant OEL. Nevertheless, it is important that the hazards are recognised, and that potentially dangerous situations and environments are identified.   There is very little toxicological information available relating to trace components of landfill gas and even less is known about the influence of these compounds on living systems when they occur in mixed phases (Scott, 1990). An assessment of the toxicological hazard posed by landfill gas is therefore difficult, and although some maximum TLV values have been derived for particular body organs to individual compounds (Scott et al, 1988), these preliminary findings did not account for potential additive, antagonistic or synergistic effects. Without such data, the process of risk evaluation (stage 3 of the risk assessment process described above) cannot be effectively undertaken.

In a study conducted by Scott (1990) only ten of one hundred and thirty six compounds detected in landfill gas were present at concentrations above their relevant TLV's or OES's at any time during the study period. This work (Scott, 1990), also suggested that any potential health risk posed by landfill gas would be likely to be associated with the period immediately following waste emplacement. Also, according to Hoather and Wright (1989), the most significant trace components are hydrogen sulphide, vinyl chloride, benzene, toluene, trichloroethane, methyl mercaptan and ethyl mercaptan, all of which have been found at concentrations above their TLV's.

However, Scott *et al*, (1988) have shown that the concentrations of the most hazardous compounds detected at a number of landfill sites could be lowered below the relevant TLV by dilution within the first metre above the landfill surface. They concluded (Scott *et al*, 1988) that "landfill gas is not expected to give rise to toxic concentrations of individual components in unconfined localities where moderate dilution is available". The potential effects of synergistic action of the various trace components could not be quantified, and thus although trace toxic compounds within landfill gas constitute a potential threat to the health of workers involved in landfill operations, and to visitors and those living and working in areas adjacent to landfill operations, the risk associated with the trace components of landfill gas cannot be quantified until further relevant data are made available, but would be likely to be low where significant dilution occurs. It is the presence of certain trace components that confer the characteristic malodour to landfill gas and which as a consequence elicits a large number of complaints to both landfill operators and regulators. This form of environmental pollution may not necessarily represent a health risk but certainly impacts on the quality of life for those affected.

Some common sources of trace compounds, their generation within landfill, and implications for landfill management are discussed in Gendebien *et al* (1992).

Global Warming Global warming or the greenhouse effect is the warming of the Earth's atmosphere caused by the accumulation within it of gases that absorb reflected solar radiation. These gases include methane, carbon dioxide, NOx, SOx, and chlorofluorocarbons. The potential long-term consequences of global warming include climate change, sea level rises (as a result of melting of polar ice) desertification, and changing food production. ETSU (1991) have estimated that landfill gas contributes 20% of atmospheric methane emissions within the UK and current estimates suggest that landfill methane contributes 1-2% to the annual rate of increase in radiation due to the accumulation of all "greenhouse gases" (Augenstein, 1990). Although there is some uncertainty regarding these figures, Blake and Rowland (1988) have shown a steady increase in atmospheric methane concentrations over the period 1978-1988.

Recent estimates by The Intergovernmental Panel on Climate Change (IPCC, 1992), suggest that landfill contributes at 30-70 Tg methane.yr[1], and that this represents 8-20% of the total anthropogenic emissions.

Without more effective integrated waste management, in which waste minimisation and recycling are encouraged, world population increases, especially in the developing countries, are likely to add further greenhouse gases to the already overburdened atmosphere. The affect of carbon dioxide, as a greenhouse gas, is variously considered as being a factor of 20-30 less damaging than methane. Thus any process which burns methane to carbon dioxide will have a significantly smaller damaging affect on the environment, and will represent a more sustainable means of waste management.

4.3   Landfill gas migration

For gas migration to occur, there must either be a concentration gradient to allow diffusion in the gaseous phase (diffusive flow), a pressure gradient (viscous flow) or a combination of both. The rate of diffusion for a gas is inversely proportional to the square root of its density. Thus a "light" gas such as methane will migrate 1.65 times faster than carbon dioxide, which is heavier (Williams and Hitchman, 1989). When a suitable migration pathway is present (e.g. fractured geological strata, mine shaft), then the gas may migrate large distances where it may affect a sensitive receptor. At sites lacking gas control measures, landfill gas has migrated 300 to 400 metres beyond the site (Esmaili, 1975). If the receptor is sensitive to the gas, then harm may occur.

During landfill development, any gas produced will vent via the pathway of least resistance. Therefore prior to final capping and assuming that only permeable intermediate cover has been used, most gas produced will vent to atmosphere. After final capping, gas venting to the atmosphere will be limited according to the effectiveness of the cap. As a result, gas pressure will develop within the landfill creating a driving force for gas migration. Under these conditions, the possibility of the lateral migration of gas increases and the rate and extent of migration will depend upon a number of factors including environmental, climatic and geophysical conditions.

Environmental factors are essentially restricted to conditions within the waste and will affect the rate and extent of waste degradation and hence affect the rate and extent of landfill gas pressure build up. According to the UK DoE (1992a), the method of filling is the most important factor affecting within site gas migration. Thin layer techniques utilising good compaction and daily cover will tend to encourage lateral gas migration, especially when low permeability cover materials are used. Conversely, the construction of wells within the site will tend to favour vertical gas migration within and around these structures. Geophysical conditions surrounding the landfill will affect the gas migration pathways; faulted and fractured strata and strata of varying gas permeabilites will affect the direction and rate of gas movement and can be modified by hydrogeological factors such as water table levels. In varied lithologies, gas will tend to migrate preferentially through beds of rock whose grain size, shape and packing are such as to make them most permeable (DoE, 1992a). Gas may also migrate through caves and cavities, fissures, mineshafts, sewers, drains, tunnels, and other features that create a path of least resistance to gas movement.

Climatic conditions including atmospheric pressure and rainfall can also affect landfill gas migration: as atmospheric pressure falls, the surface pressures opposing gas migration decrease, thus facilitating gas movement from the landfill. The pressure differential between the landfill gas and atmospheric pressure is therefore important, and an inverse relationship between atmospheric pressure and gas migration (measured as methane concentration at off-site monitoring points) can be demonstrated at a number of landfill sites. Using computer models, it has been predicted that the increase in gas emissions from a landfill is proportional to the rate at which surface pressure is changing, rather than the actual value of surface pressure *per se*. Thus gas emissions are highest during times of rapid atmospheric pressure decrease (DoE, 1993c). The same model also predicts that it is the proportion of open void space in the ground, rather than the permeability which determines the variability of gas emissions. Thus gases monitored in clays will exhibit greater variability than in sandstones, even though the underlying gas regime is the same. Other model predictions include:

- landfill gas composition is not only influenced by instantaneous pressure changes, but the weather pattern over the preceeding days and weeks
- an ideal monitoring programme would involve continuous monitoring of pressure, and would cover a period of several rapid rises and falls in atmospheric pressure
- that variations in landfill gas composition are greatest nearest to the surface of the landfill

When the model was tested against actual data collected from a landfill, it showed that the size of variations was greatest nearest to the surface, and that the nature of the variations reflected preceeding atmospheric events (DoE, 1993c). A rapid fall in atmospheric pressure is thought to be responsible for the house explosion at Loscoe, Derbyshire in 1986 (Williams and Aitkenhead, 1991). The model above predicted an increase in methane content of the gas, and that the large fall in barometric pressure which occurred on the day of the explosion would have caused a five-fold increase in gas flux at the outcrop of the permeable sandstone through which it migrated. Models such as that described are in the early stages of development, and are reliant upon the quality of data fed into the model, but may find increasing use in the long-term monitoring and control of closed landfill sites.

The migration of landfill gas can also be affected by rainfall, as it will cause surface materials within landfill caps to swell and close surface cracks, thus reducing vertical migration pathways, with a resultant increase in potential for lateral gas migration. Water infiltration can also increase water table levels outside the landfill and leachate levels within the site, thus reducing the gas volume and increasing gas pressure.

Thus it can be seen that the rate and extent of landfill gas migration is affected by a wide range of factors that are essentially physical in nature. Although biological and chemical processes may not significantly affect the volume or rate of migration, they may alter the gas composition, through processes such as methane oxidation which converts methane to carbon dioxide. Factors affecting the migration of landfill gas have been considered in more detail by Campbell (1989; 1991a; 1991b). The prediction and evaluation of landfill gas impacts have been considered by Petts and Eduljee (1994).

4.4   Control of landfill gas

Incidents such as those described above, have highlighted the dangers associated with landfill gas and serve to emphasise the need for effective control of these emissions. This control can be effected either through controlling waste inputs (to prevent or limit emplacement of organic waste), through controlling the processes within the waste (e.g. minimising moisture content to limit gas production), or through control of the migration process (by the use of either physical barriers to migration, or by reducing the motive force, through the use of 'vents' to remove gas from the site and hence reduce gas pressure). The potential, and relative merits of the first two options, and their relationship to the sustainable landfill are discussed in Chapter 7.

Physical barriers will be of limited use in the prevention of gas migration, and are seldom used as the sole means of gas control. Such barriers do not limit gas production, and as the gas must migrate from the point of generation, they can only be used to exert limited control on the *path* of gas migration. In this way, slurry cut-off walls, geomembranes, or similar, may limit migration in the direction of a particularly sensitive receptor (e.g. building), but will increase vertical gas migration. Natural clays can be used alone or in conjunction with synthetic lining materials such as HDPE flexible membrane liners to form basal and side-wall lining to landfills but the prime function of such systems is the control of leachate migration. However, even under ideal conditions, natural clays are relatively permeable to gas migration (Figueroa and Stegmann, 1991) and will not **prevent** gas flow. If natural clays are allowed to dry to the extent that cracks develop, then gas movement will increase accordingly.

While synthetic landfill liners constructed of materials such as HDPE will be more effective in controlling lateral gas migration (but not to the extent that they could be used as a sole source of gas control) they will not be as effective in preventing gas emissions to atmosphere through the landfill caps. When used within caps, the primary aim of geomembranes is to prevent water infiltration; adjoining sheets are often overlapped rather than sealed and thus facilitate gas flow, while ensuring run-off of surface liquids. If the cap seal was gas-tight then gas pressure within the the site would increase and thus increase the possibility of lateral gas migration. Furthermore, geomembranes cannot be fitted retrospectively, and therefore, except when used in landfill caps, cannot be used for gas control in the majority of landfills that are either closed or already operational but which do not use such protection.

Because gas migration cannot be easily prevented, removal of the landfill gas, to reduce the driving force for gas movement is often the preferred option. This is normally undertaken by the use of within waste vents (wells) through which the gas can be extracted either actively or passively, or by the use of stone-filled vent trenches, often placed around the periphery of landfill sites. The design of a typical gas vent system is shown in Figure 9.

4.4.1 Passive control
The design of typical passive gas collection systems can vary. Gas wells (commonly drilled vertically into the waste, after filling to final levels) placed within the waste act as channels for gas venting to the atmosphere. The rate of gas flow will be dependent upon the physical waste characteristics, gas generation within the waste, the associated gas pressure and the atmospheric pressure. At constant gas generation rates, the amount of gas venting could be anticipated to show an approximately inverse relationship to atmospheric pressure (but see above). The amount of gas abstracted by passive systems will be less than that using an active abstraction system and this factor together with the fluctuating gas concentration, make the collection and flaring of the gas at a dedicated landfill gas flare difficult, and for this reason gas flaring is seldom used with passive control systems. Thus although in many circumstances these simple and relatively inexpensive systems may effectively control lateral gas migration, and reduce the off-site hazard, the surface landfill gas emissions will increase, with a resultant increase in the release of "greenhouse gases" and toxic trace components. Passive venting systems are unlikely to be effective when used alone for the control of gas migration in sites with high rates of gas generation, and recent draft guidance in the UK (DoE, 1995b) recommends that they should only be used in situations where the rate of gas generation is low.

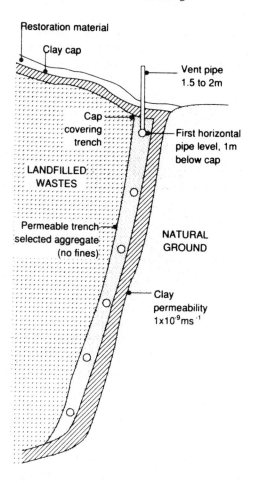

Restoration material

Clay cap

Vent pipe
1.5 to 2m

Cap
covering
trench

First horizontal
pipe level, 1m
below cap

LANDFILLED
WASTES

Permeable trench
selected aggregate
(no fines)

NATURAL
GROUND

Clay
permeability
$1 \times 10^{-9} ms^{-1}$

Figure 9.   Permeable trench venting system (DoE, 1992)

Stone-filled vent trenches represent a second form of passive gas control. These trenches are often used on site boundaries where gas migration has been identified, but such systems can only be effectively used when the landfill depth does not exceed 8m (DoE, 1992a), and they are seldom used at depths greater than 5m. The limit on the effective depth of trench is controlled by the depth to which an excavator may dig. To excavate deeper than this requires specialised equipment or techniques and is not cost-effective. The effectiveness of vent trenches (Figure 8) increases when they are used in conjunction with a geomembrane, placed within the trench along the wall furthest from the gas source. The vent trenches again offer a path of least resistance to gas flow, but increase surface emissions at the expense of lateral gas migration. Careful management of these trenches is required to ensure that venting surfaces do not become blocked.

For both passive vent wells and vent trenches there is potential for air ingress through the system, especially at times of high atmospheric pressure. This may result in an increase of oxygen levels within the waste and an associated potential for fire. To minimise this risk, vent trenches are often sealed at the surface, and constructed with vertical risers through which the gas can vent. In this way, the surface area through which air can enter the landfill is minimised without affecting the performance of the trench as a landfill gas venting mechanism. Water ingress which may increase leachate levels within the waste will also be minimised. Advice on the construction of vent trenches is given in Waste Management Paper No 27 (DoE, 1992a).

4.4.2 Active Control

Active gas control systems utilise energy to "pull" the gas from the landfill waste. The gas collection systems usually comprise an array of interconnected vertical, or in some cases, horizontal, perforated pipes within the waste, through which gas is abstracted. The systems described under "passive control" (above) may all be modified for active gas abstraction. The collected gas may be either flared, or if sufficient gas is present and the economics are viable, may be burned as a direct heating fuel or as a fuel in electricity generation.

The advantages of an active gas control system are that the abstraction can be controlled, the reduction of within-waste gas pressure will be more effective than with passive systems, and because in most situations, the collected gas is burned, the atmospheric pollution potential is reduced. Although gas flaring does not remove all trace components and bulk gases, the concentrations of each are significantly reduced.

In exteme cases, lateral gas migration may continue after the installation of within-site gas control measures and may require gas abstraction off-site, along the path of the migrating gas. The effectiveness of this form of control varies in a site specific manner according to local conditions. However gas control can also be exercised at the target under threat. In the case of landfill gas, the 'target' is usually a building, in which gas can accumulate within a confined space and which is therefore at risk from explosion. Gas control at the 'target' can be either active, passive or semi-passive and coupled with effective building design can reduce the explosive hazard in a way which can facilitate safe building construction both close to and on top of closed landfill sites.

4.4.3   Gas control for buildings protection

When migrating landfill gas collects in enclosed spaces within buildings, relatively high flammable gas concentrations can be reached locally, creating highly dangerous circumstances where fire and explosions may occur. There are many recorded examples where landfill gas has been responsible for explosions within buildings (e.g. Gendebien *et al*, 1992). For a landfill gas - related fire or explosion to occur, three fundamental conditions must be satisfied.

*   Flammable gas concentrations must be within the flammable range
*   Oxygen must be present
*   There must be a source of ignition

As oxygen is a component of air, removal of oxygen does not represent a practicable option. Also, as, for example, the spark generated within light switches is sufficient to provide the source of ignition, the removal of the source of ignition can be difficult to achieve and maintain in all buildings.   Thus in most circumstances, buildings protection against landfill gas - related explosions is achieved by gas control to maintain flammable gas levels below the lower explosive limit (LEL).

The means by which gas control is achieved is dependent upon the local conditions. For example, the type and extent of controls for buildings at some distance from a landfill are likely to be quite different from those in buildings constructed on the site of a closed landfill.

Before addressing the types of control that may be applied, it is worth considering the issues associated with construction on former landfill sites: The main reason that companies wish to construct buildings on closed landfills is that the cost of the land is relatively cheap. For some developments, the use of closed landfills may be appropriate, but for others the associated risks may be too high. In the latter category, it is difficult to foresee **any** circumstances under which the development of housing could be supported; In any proposed development an appropriate risk assessment should have been conducted, and where the consequences of fire or explosion (and other gas related hazards) are so severe, then even uncertainties relating to the nature of emplaced waste are likely to be enough to preclude development. Furthermore, even if the hazards are controlled through the use of barriers (including geomembranes) and other mechanisms, difficulties in controlling the activities of residents and ensuring that actions of the householders do not compromise the control systems (especially in the long-term) are also likely to preclude any housing construction. Despite the obvious risks, there are many examples of housing development on closed landfill sites. It is to be hoped that such development is prevented in the future.

However, under the right circumstances (where the associated risks have been deemed to be acceptable) some development on landfill sites may be possible.

Many construction programmes on former landfill sites have been undertaken by companies requiring large car parks, where the building is constructed on virgin land adjacent to the landfill, and the landfill area is converted to a car park (e.g. superstores). In other cases, warehousing and light industrial units have been constructed directly on the landfill. There is not scope here to discuss the circumstances under which construction on closed landfills may and may not be appropriate. However, factors that will influence the decision include:

- the nature of the emplaced waste (biodegradable? toxic? etc.)
- the depth of fill (this will affect the gas production potential and long-term settlement which will, in turn, affect foundations and construction techniques (e.g. piled/raft construction)
- the results of long-term monitoring programmes (e.g. gas concentration, leachate composition/depth)
- the nature of the surrounding geology
- the proximity and sensitivity of "receptors"
- the type of proposed development

This list shows that a lot of relevant information must be gathered over a period of time before an appropriately informed decision can be made. Once this information has been collected and assessed then the most effective gas control systems can be designed and implemented. Gas control may be achieved both beyond the building and within the building. Gas control beyond the building may be either active or passive and will utilise the types of control described above.

Gas control within buildings can be achieved by a variety of means. The major principle adopted in building protection is that of construction upon gas-impermeable foundations. To be successful, this method requires that the impermeable barrier remains intact. This can be difficult to achieve as many mains services enter buildings through the floor slab, and floor joints which should always be effectively sealed, are susceptible to the effects of differential settlement. However, problems associated with the above can be effectively dealt with and there are many examples of successful development on landfill sites.

The most common means of constructing a gas-impermeable floor slab is to install a membrane liner. In many cases this may be all that is required while in circumstances where gas related hazards are higher, underslab venting may also be used.

The key requirement is that the membrane must remain intact both during construction and during the lifetime of the building. Quality control is an important factor in ensuring the above, while building design to allow access of mains services at a point above the floor slab will also be important. Any tears during the construction phase must be effectively sealed and tested to ensure integrity.

Under-floor voids are often used to aid gas control, whereby the under-floor flow of air, which may be either actively or passively undertaken, is used to dilute gas entering the void. The most effective use of this method would also require the use of an impermeable membrane barrier below the void, as described above.

Many systems rely on "passive" air flow in which the pressure differential across an underflow void induces air flow. Semi-passive systems utilise rotating cowels, or similar devices to induce air flow as wind flows over them. Active gas control is carried out by means of gas blowers which may be linked to gas sensors and which will 'cut-in' when a lower gas level is exceeded. If the fans are ineffective, and an upper gas level is subsequently exceeded then alarms will sound to signal building evacuation.

The automatic, remote sensors and associated display panels should be checked on a regular basis and additional manual monitoring conducted to assure the effectiveness of the automatic system. Recent models are equipped with data loggers from which stored data may be automatically transmitted to a central facility and computed to show the change in gas levels over time. Such data are often more revealing than point source data. However care must be taken in interpreting data as in many cases only flammable gas will be monitored. (Examples have been recorded where vapours from nearby road construction or roof asphalting have triggered building alarms). For effective use the gas sensors should also be placed in regions where gas concentrations could be anticipated to be highest.

Active gas abstraction systems can be particularly effective when coupled to a system of perforated pipes within a stone-filled void which will ensure gas abstraction at "susceptible" regions. Such systems should be carefully "balanced" by the use of valves or by using pipework of different diameters, to ensure active abstraction throughout the pipe network including appropriate building design (e.g. all mains services to enter building above floor slab).

Further consideration of gas control and the assessment of the design, use and effectiveness of within-building gas control measures can be found elsewhere (Card, 1992).

4.5.   Gas Monitoring

The monitoring of landfill gas is a requirement of both proposed European legislation (CEC, 1991), and of subtitle D of RCRA. In the USA, subtitle D requires a routine monitoring programme for methane, and that this should be based upon facility-specific soil conditions. If methane levels are exceeded, then the owner/operator is required to undertake a number of actions to address the problem. These are summarised in USEPA (1994). One of the major differences between US and European gas monitoring requirements is the requirement in the USA for monitoring surface emissions of non-methane organic compounds (discussed above), and that while monitoring programmes in the USA are based on assessed risk, those in Europe set down minimum requirements for all sites, and take little account of local conditions. In this respect the system as used within RCRA appears to offer many advantages over the proposed European system. Advice on gas monitoring procedures and equipment can be found elsewhere (e.g.Institute of Wastes Management, 1991; DoE, 1992a; USEPA, 1994).

4.6   Gas Utilisation

For most countries, the production of energy from landfill gas is not competitive at today's energy prices, but it has been identified as a promising renewable energy source. Within the UK, "new and renewable energy sources have the potential to make a large impact on UK electricity needs in the next century" (Anon., 1994d). Energy from landfill gas is one such renewable energy source, and in order to help support the use of renewable energy in the short term, the UK Government has introduced the Non-Fossil Fuel Obligation (NFFO) with the intention that after initial support, these schemes will be able to compete without additional aid.

The UK Department of Trade and Indstry (DTI) currently supports a landfill gas (LFG) programme with three key aims. They are (From Meadows *et al*, 1994):

- to provide the government with a thorough assessment of the potential and prospects for LFG use

- to encourage maximum exploitation of LFG where this is economically attractive and environmentally acceptable

- to encourage UK industry in developing capabilities for domestic and export markets.

"The overall objective of the LFG programme, and of NFFO support, is to increase the rate of electrical energy production from the current energy level of 0.6 TWh/y to over 3.6 TWh/y in 2005 and over 4.9 TWh/y by 2025" (Meadows *et al*, 1994).

For effective utilisation in gas engines or turbines, the methane content of landfill gas should be approximately 50%. However, where gas collection is used primarily for the control of migration and the protection of 'sensitive targets' then the methane content of the gas is often much less than 50% in order to maintain a flame at the gas flare. For this reason, it is important clearly to identify at the outset whether the gas collection system is for gas control or energy generation. Local site conditions may require the use of both types of system where, for instance, peripheral wells are used for gas migration control and central wells are used for collection with subsequent utilisation for electricity production. It is also possible for wells to be designed and built to accomodate both systems and to be switched from one to the other when the situation demands. However, the cost of such a system will be much higher than a simple system and this must be accounted for when calculating the economic feasibility.

The high moisture content of landfill gas and the presence of trace corrosive gases requires that the collected gas should be pre-treated before combustion in a gas engine. Dewatering equipment often takes the form of expansion chambers where the gas is allowed to cool and for the moisture to condense. It will also be important for gas collection pipes to be installed to ensure that condensed moisture in the pipeline is not allowed to restrict gas flows and that suitable collection facilities (known as water knockout pots) are placed at low points within the pipeline. Most gas collection systems are designed to cater for the production of approximately 3 litres of condensate per 100 cubic metres of gas.

Within the UK relatively little pre-treatment of gas, other than dewatering is undertaken. However, one site in Germany has installed a liquid scrubbing system to remove acid gases prior to combustion in spark-ignition engines to limit the formation of dioxins (Eden, 1994). Gas-scrubbing processes that may be used include water scrubbing, alkali-scrubbing, adsorption processes and refrigeration processes.

Because the amount of gas produced from a landfill can vary with time, energy utilisation systems are designed to run on less than the maximum amount of available gas. As a result, a flare is required to burn the excess gas which may contain a number of potentially toxic components.

A number of reviews of the utilisation of landfill gas have been produced including those of Richards and Aitcheson (1991), and Gendebien et al, (1992).

CHAPTER 5

LANDFILL LEACHATE

5.1   Introduction

One of the major hazards associated with landfill leachate is the potential to contaminate or pollute groundwater and surface water supplies. The USEPA has estimated that there are about 55,000 landfills in the USA, approximately 75% of which are polluting groundwaters (cited in Jones-Lee and Lee, 1993). Within Europe, equivalent figures are not available, but as indicated in Chapter 2, the pollution of groundwater by landfill leachate has been recognised. The degree of environmental pollution caused by leachate will be dependent upon a range of factors including local geology and hydrogeology, the nature of the infilled waste, and the proximity of susceptible receptors.

The dependence on groundwater for drinking water supplies will vary from country to country and region to region, but irrespective of the local conditions, the supply will be at threat from a range of potential pollution sources, including landfill. Groundwater flow rates can be very slow and are often measured in metres or tens of metres per year. As a result, an aquifer may only be identified as polluted long after the pollution event has occurred, at which time remediation may not be possible, or if possible, will be very expensive. Such pollution events are in general often difficult, and when possible, very expensive to remediate and although both the unsaturated (vadose) zone and saturated (groundwater) zone have potential to attenuate some pollutants, pollution **prevention** is paramount in the protection of the environment. In this context pollution and contamination must be differentiated. According to the tenth report of the Royal Commission on Environmental Pollution (1984), pollution is defined as "the introduction by man into the environment of substances or energy liable to cause hazards to human health, harm to living resources and ecological systems, damage to structures or amenity, or interference with legitimate uses of the environment". For the purpose of this discussion "contamination" represents the introduction by man into the environment of potentially dangerous substances which do not **necessarily** constitute a hazard. Where a hazard is present the term "pollution" is used.

Within the UK, the recently published Groundwater Protection Policy (NRA, 1992) identifies the framework to be used in controlling groundwater pollution from a number of potential sources, including landfill. These controls supplement those of the European Groundwater Directive. In the USA, control of leachate from MSW landfills is exercised under sub-title D of RCRA, which also places controls on the levels of contaminants that must not be exceeded in the uppermost aquifer at a point which should be no more than 150 metres from the landfill boundary (Table 10). These controls serve as examples of the increasing national legislation throughout much of the developed world aimed at reducing environmental pollution, which places ever stricter controls on the design, management, and operation of potentially polluting industries. However, such controls have not always been in place, and mention has already been made of environmental pollution incidents that have occurred as a result of landfilling activities.

As long as landfill continues as an accepted form of waste disposal, there will always be potential for related pollution events, and an understanding of the factors affecting such pollution is important.

5.2   Leachate Production

Landfill leachate is comprised of the soluble components of waste and the soluble intermediates and products of waste degradation which enter water as it percolates through the waste body. The processes and products of waste degradation have been shown in Figure 5 and show the close relationship between landfill gas and landfill leachate production.

The amount of leachate generated is dependent upon a number of factors which can be summarised as follows:

•   water availability

•   landfill surface conditions

•   refuse state

•   conditions in the surrounding strata

Table 10.   **Maximum Contaminant Levels (MCLs)**
Point-of-Compliance Performance-Based Criteria

Chemical	MCLs (mg/L)
Arsenic	0.05
Barium	1.0
Benzene	0.005
Cadmium	0.01
Carbon tetrachloride	0.005
Chromium (hexavalent)	0.05
2,4-Dichlorophenoxy acetic acid	0.1
1,4-Dichlorobenzene	0.075
1,2-Dichloroethane	0.005
1,1-Dichloroethylene	0.007
Endrin	0.0002
Fluoride	4.0
Lindane	0.004
Lead	0.05
Mercury	0.002
Methoxychlor	0.1
Nitrate	10.0
Selenium	0.01
Silver	0.05
Toxaphene	0.005
1,1,1-Trichloromethane	0.2
Trichloroethylene	0.005
2,4,5-Trichlorophenoxy acetic acid	0.01
Vinyl chloride	0.002

Source: *Federal Register*, October 9, 1991 (40 CFR Part 258.40)

Measures of each of the above determine the water balance equation, a simplified version of which appears in the UK Department of Environment Waste Management Paper 26 (1986):

$$Lo = 1 - E - aW$$

where

Lo =	Free Leachate retained at site (equivalent to leachate production - leachate leaving the site)
I =	Total liquid input
E =	Evapotranspiration losses
a =	Absorptive capacity of the waste
W =	Weight of waste deposited

Good landfill practice normally requires that Lo is negative or zero and therefore that no excess leachate is produced

i.e. $I - E < aW$

A predicted unfavourable water balance (net liquid production) at the design stage would require the selection of an alternative site or the redesign of engineering and operational parameters in a way that would reduce the input (I) or increase the output (E) in the above equation, and thus reduce the amount of liquid arising within the landfill.

The factors affecting water availability include precipitation, surface run-off, groundwater intrusion, irrigation, liquid waste disposal and refuse decomposition. Surface run-off, groundwater intrusion and irrigation can be controlled through effective site design and operation. Refuse decomposition may also be controlled to some extent inasmuch as the major factors affecting waste decomposition have been identified (Section 2.3) and can be optimised where practical and economically feasible. However, precipitation is the primary contributor to the water balance equation (Fenn et al, 1975; Dass et al, 1977) and cannot be controlled. Rainfall figures must be used therefore at the site design stage in order to ensure a net water deficit.

Surface conditions which may affect leachate generation include vegetation, cover material (density, permeability, moisture content etc.), surface topography and local meteorological conditions. The latter factor may affect moisture conditions in the cover material both directly and through the effect on evaporation and transpiration (evapotranspiration) processes of the surface vegetation.

The refuse state will affect the "field capacity" of the waste; "field capacity" being defined as "the maximum moisture content which a soil or a solid material can retain in a gravitational field without producing continuous downward percolation" (Lu *et al* 1985). The water content of the waste will be affected by a number of factors including the water content at emplacement, infiltration during landfill development and the waste composition and density.

The "conditions in the surrounding strata" will encompass water table levels where, for example, a high water table or underground spring will have a greater impact on leachate generation when the site is "sealed" with relatively permeable (mineral) liners rather than the much less permeable synthetic liners, although in the latter case engineering design and controls must ensure the long term effectiveness and integrity of the synthetic liners. Significant groundwater influx at the construction stage can create many problems (such as destabilisation of the surrounding natural ground), and in many cases such problems would favour selection of alternative, more appropriate sites.

Each of the above factors can be measured with a varying degree of accuracy and inserted into a water balance equation, which will account for the total liquid inputs and total water leaving the landfill. A more detailed review of the relevant factors has been conducted by Knox (1991).

5.3   Leachate composition and properties

The composition of landfill leachate may be expected to vary with time, and from site to site, as a consequence of variable water infiltration rates and amounts, differing ages of sites, differing waste composition and local environmental conditions. The processes of leachate generation within the site have already been discussed, and the change in leachate composition as the waste moves from the acetogenic phase to the methanogenic phase have been described briefly.

Domestic refuse would normally produce leachate with the highest BOD (high strength leachate) but may be low in individual hazardous components. However, even domestic refuse will not be free of hazardous materials. According to Jones-Lee and Lee (1993), it has been estimated that each person (in the USA) contributes 4 litres.yr[-1] of hazardous chemicals to their household waste stream. These chemicals can include pesticides, paint residues, and mercury from fluorescent tubes and batteries. Commercial waste may also produce a high strength leachate, while waste classified as "inert" is, in effect, seldom inert and often contains a proportion of degradable components which may lead to the production of a leachate capable of pollution.

It is difficult, because of the above, to describe a typical landfill leachate, although most leachates will contain characteristic components which change as the waste moves from the acetogenic phase to the methanogenic phase. However the data in Table 6 are, within the confines of the above, considered to be typical acetogenic and methanogenic leachates. In relation to Table 6 the following points are worthy of note:

pH
Methanogenic bacteria utilise available hydrogen, affecting bacterial growth in a way which favours production of acetate rather than higher acids (e.g. propionate, butyrate). The net effect is to ultimately raise the pH. As long as methanogenic activity continues, this will act to maintain the pH near neutrality.

Chloride
The chloride ion is a so-called conservative ion, it is relatively unreactive, and will undergo little or no retention as it moves through waste. It acts as an indicator of dilution, and can be used to map the leading edge of a pollution plume.

Organic-Nitrogen
The decrease in organic nitrogen in methanogenic waste is a function of the microbial degradation of compounds such as protein. The degradation of protein appears to occur rapidly during the acetogenic phase, resulting in the production of high levels of ammonia (present as the ammonium ion). Ammoniacal nitrogen and organic nitrogen are differentiated.

Ammoniacal-N
Ammoniacal nitrogen is probably the most important inorganic component and will often be present at high levels even in methanogenic waste, and because of this, it is often necessary for leachate to be treated off-site before discharge to sewer. Once released within a landfill, there is no significant biochemical pathway whereby ammonia can leave the waste (Robinson and Gronow, 1992), and thus controlled treatment is required to remove this component.

TOC, COD, BOD
These parameters are measures of the amount of carbonaceous material present. For each, there is a significant reduction in the methanogenic waste. An important factor in this is the conversion of organic acids to methane and carbon dioxide thereby decreasing the concentration of organic carbon in the leachate. It has been shown (Pohland, 1975) that these acids can represent the bulk of the organic carbon present in leachate during the early stages of waste degradation. Further work by Belevi and Baccini (1989) showed that more than 22% of the initial carbon content of MSW was exported as gas within 10 years of waste deposition, while with increasing waste age humic and fulvic-like compounds become more predominant (Chian and DeWalle, 1977). Conversion of organic carbon within leachate to the various components of landfill gas will reduce the leachate associated hazard while increasing the landfill gas-associated hazard.

BOD/COD ratio

This ratio is an indication of the proportion of biologically degradable carbon to "total" (chemically oxidisable) carbon and will decrease in the methanogenic waste as the amount of degradable carbon decreases as a result of microbial activity.

Landfill leachate is hazardous by virtue of the high BOD and COD content and the toxicity of dissolved organic and inorganic chemicals, and according to Harris (1988), present day domestic waste has a much higher pollution potential than waste deposited prior to the 1960's. The increased pollution potential will be related to the changing waste composition in recent years, discussed previously in Chapter 2, and factors such as the increase in total waste arisings. Dissolved landfill gas will also create a potential problem especially if it can accumulate in a confined space after moving out of solution. Degassing requirements and relevant safety levels have been dicussed further by the North West Waste Disposal Officers (1991).

If allowed to enter streams and other watercourses, aerobic metabolism by plant life and microorganisms, of the compounds that make up the BOD and COD, can remove the majority of dissolved oxygen present within the water, converting the watercourse to a eutrophic state in which fish and other animal life cannot survive.

Hazards and associated risks due to other organic and inorganic solutes will vary according to the nature of the leachate, but the heavy metals and ammonia which are present in most leachates are recognised as common toxic components, and even when present at relatively low levels, heavy metals may prove hazardous if they are bio-accumulated through the feeding chain.

5.4   Leachate migration

The importance of pollution prevention through hazard reduction and containment has been emphasised in previous chapters, for once leachate escapes to the surrounding environment effective control is lost, as discussed previously, remediation can be too difficult, too expensive, or both for effective protection of sensitive receptors to be ensured. However, as part of a risk management process, it is important to understand the factors that control the fate of leachate upon escape. Our knowledge of hydraulics and attenuation factors is increasing, but because of the complex nature of the sub-surface region, accurate prediction of leachate fate is, in most circumstances, impossible.

Upon escape, landfill leachate will move into the surrounding environment in a way that will be determined by a range of factors, the most important of which will be the nature of the surrounding geology and hydrogeology. Escape to surface waters may be relatively easily controlled and the pollutant fate will vary according to the nature of the receiving waters. Escape to groundwater may be much more difficult to control and will almost certainly be more difficult to clean-up. In this case, the pollutant fate will be dependent upon a range of factors associated with the nature of the various phases of the sub-surface region.

As described earlier, the sub-surface region of the earth comprises the unsaturated zone and the saturated zone. The boundary between the two zones is represented by the water table, and the region immediately above the water table is the capillary zone, where pore spaces within the unsaturated zone may be saturated by water rising from the water table under capillary forces.

The understanding of pollutant fate and transport in the sub-surface environment is an important factor in successful groundwater quality management. Once landfill leachate has escaped into the sub-surface environment, the degree of contamination of groundwater will be affected by physical, chemical and biological actions that act in a similar way to those within the landfill, as described earlier. The relative importance of the individual processes may change however as a result of the change of environment from that of the landfill to the sub-surface region.

At a fundamental level, groundwater contaminants are affected by two opposing actions: the tendency to be entrained and transported in the body of groundwater, and the tendency for solutes to be attenuated by various reactions en route through the aquifer.

5.4.1 Physico-chemical processes affecting leachate migration and attenuation

Physical dispersion, dilution and molecular diffusion tend to decrease pollutant concentration at any point, and dilution effects will be determined by the rate of water flow in both the saturated and unsaturated zones. Water flow in the unsaturated zone is essentially vertically downward under the force of gravity, whereas in the saturated zone, flow direction depends on the groundwater flow field, which is determined by factors including hydraulic head and conductivity, water sources and sinks and the characteristics of the aquifer including its boundary conditions. For landfill leachate, dilution necessarily contaminates the relatively clean diluting groundwater, and is recognised as one of the attenuating factors in the dilute and attenuate approach to landfill (Chapter 2).

The above processes are essentially physical in nature and mainly occur within the saturated zone. However, leachate will not always escape directly into groundwater and so it is also necessary to consider movement in the unsaturated zone where there may be greater potential for chemical and biological attenuation.

The unsaturated zone is often the first line of natural defence against groundwater pollution, and because water movement through it is slow and confined to smaller pores with larger surface areas, it provides a very favourable environment for pollutant attenuation or limitation (Foster, 1987) This attenuation in the unsaturated zone represents a key element of the dilute and attenuate landfill (Chapter 2). The effectiveness of the attenuation (which as a natural process cannot be managed) will vary according to the leachate composition and local conditions. For this reason, dilute and attenuate landfill designs should only proceed on an assessed risk basis. The attenuation processes are described briefly below. Whether discussing the saturated or unsaturated regions, the types of attenuation processes will be similar.

According to Gera (1988), the main causes of retardation of substances in groundwater are ion-exchange, precipitation and dissolution, generation of insoluble complexes and the generation of colloids, followed by flocculation and filtration. Many of these will be affected by conditions of pH, temperature and redox potential, amongst others. The cumulative affect of these chemical interactions will be to limit or increase the activity and movement of pollutants in groundwater. Adsorption, ion-exchange and precipitation tend to restrict leachate movement through substrata, while complexation of heavy metals with organic material may facilitate transport of the metals through aquifers. These processes have been described briefly in Chapter 3. In laboratory studies, Christensen *et al* (1989), showed that most pollutants tested were attenuated by a combination of sorption and precipitation reactions. The movement of zinc and cadmium was especially restricted in the anaerobic zone of even very coarse aquifer materials.

Most leachates can be assumed to have much lower pH values and redox potentials than the ambient groundwater; dispersion therefore causes continual mixing of waters that are different in chemical composition and redox status, and a cycle of changing Eh and pH causes reactions to take place, which in turn stimulate further changes in Eh and pH, and so the cycle continues (Cherry *et al*, 1984).

At any one time it should be possible to identify a series of redox zones in the groundwater in which different types of physico-chemical reaction can take place. Close to the leakage point a strongly anaerobic zone will develop with a redox potential suitable for methane generation and various redox processes such as sulphate, iron, manganese and nitrate reduction. Greatly diluted and attenuated leachate will eventually enter an aerobic zone where free dissolved oxygen is present in modest concentrations (Christensen *et al*, 1989). Between the two extremes, a region of intermediate Eh and pH conditions will exist and which will facilitate a greater range of physico-chemical reactions, and which may thus facilitate more effective attenuation of landfill leachate as it flows from a landfill.

5.4.2   Biological Processes in leachate migration and attenuation

The biological processes affecting pollutants in the sub-surface environment are not as well understood, but a realisation of the ability of micro-organisms to degrade many pollutants has stimulated research into this aspect of groundwater pollution. For example, laboratory studies by Christensen *et al* (1989) have shown that while the volatile fatty acid (VFA) component of the COD readily underwent anaerobic microbiological degradation, the remainder of the COD did not, and in the aerobic zone only the chlorinated aliphatics, of the 22 compounds studied showed no sign of degradation within the 280 day study. Microbiologically transformed pollutants may also be more or less toxic, and can have an important effect on the nature and extent of leachate pollution of groundwater.   The importance of biological attenuation, and the fate of some of the major pollutant groups in the sub-surface region is considered below.

Nitrogen

Dissolved nitrogen can be present in many forms although those with a major impact on the environment are ammonium and nitrate. The important biological transformations affecting nitrogen are ammonification (in which ammoniacal nitrogen is released from nitrogen-containing organic compounds), and nitrification, (where ammonium is oxidised to nitrate). Both these processes usually occur above the water table, and generally in the soil zone where organic matter and oxygen are abundant (Freeze and Cherry, 1979).

Ammonium may be discharged directly into the sub-surface, or can be generated within the soil by ammonification. Its transport and fate involves a combination of the processes of adsorption, cation exchange, incorporation into biomass and release to the atmosphere in a gaseous form (Canter *et al*, 1987). Adsorption is possibly the main method of ammonium removal, and is particularly important under anaerobic conditions. With increasing pH, ammonium reacts to produce ammonia gas and can therefore be more easily released from the soil.

The negative valency of nitrate means that under normal soil and groundwater conditions there is little attraction to soil particles. Nitrates are therefore very mobile and can move through the sub-surface with minimal transformation and retardation (Canter *et al*, 1987). It is therefore important to control, as much as possible, the levels of nitrate either **within** the landfill, or by managed collection and treatment. Also, for most landfills with low infiltration rates, it may take hundreds of years for ammoniacal nitrogen to reach levels at which leachate can be discharged to surface waters (Robinson, 1990b). Thus for containment landfills, leachate treatment to remove ammoniacal nitrogen and nitrate will be a prime requirement prior to discharge.

Metals

The attenuation of metals by sorption (especially through ion-exchange) and precipitation considerably limits groundwater pollution by these leachate components. The four major metal attenuation processes in the sub-surface region are adsorption, ion-exchange, precipitation and complexation with organic material (Canter *et al*, 1987). The important metals in terms of environmental hazards are the heavy metals and adsorption is probably the most important process of fixation and retardation for these metals. In some geologic materials trace-metal adsorption is controlled by substances which are present in only small quantities. For example the principal control on the fixation of Co, Ni, Cu and Zn in soils and freshwater sediments is the presence of hydrous oxides of Fe and Mn (Jenne, 1968; Freeze and Cherry, 1979). Furthermore, according to Lu *et al* (1985) Cd, Cu, Pb, Ni and Zn can form carbonates in relatively oxidising environments and sulphides in reducing environments where soluble sulphide is present; these processes are more pronounced in neutral and alkaline environments, while in acidic environments adsorption will usually prevail.

Ion exchange processes are limited by the competing effects of alkali earth metals such as Ca, Na and K (Canter *et al*, 1987), and usually occur on silicate clays and organic matter. Precipitation reactions are greatly influenced by pH, and are more common at neutral to high pH values and high redox-potentials (oxidised metal species are usually less soluble than reduced varieties (Canter *et al*, 1987)). For some metals therefore mobility will be increased in the anaerobic conditions beneath landfill. However, such conditions promote the formation of relatively insoluble sulphide minerals, a process which can limit trace metals to extremely low concentrations (Freeze and Cherry, 1979), and in recent studies of a migrating leachate plume (Williams *et al*, 1991), the precipitation of metals as insoluble sulphides was an important attenuating mechanism.

In recent studies (Andersen *et al*, 1991), stable metal complexes have been shown to account for a substantial fraction of the observed metal concentrations in landfill leachate pollution plume and in this case heavy metals did not pose a threat to the environment.

Organic Substances

The major processes which limit the mobility of organic substances are chemical precipitation, chemical degradation, volatilisation, biological degradation, biological uptake and adsorption (Canter *et al*, 1987). Many organic substances have extremely low solubilities in water which limits the possibility for appreciable migration in groundwater. However, many of these substances are toxic at very low concentrations, and therefore solubility constraints are therefore not capable of totally preventing migration at significant concentration levels (Freeze and Cherry, 1979).

Sorption processes appear to be of only minor significance in the attenuation of organics, where adsorption of organic compounds occurs primarily on particulate organic matter in sediments (Cherry *et al*, 1984). The structure and effective surface area of the organic matter are therefore significant variables in the retardation of organic contaminants. Of equal importance is the hydrophobicity of the contaminant.

Biodegradation of a broad range of organic compounds has been demonstrated in laboratory studies of soils, sediments and waters (Cherry *et al*, 1984), while the effect of biodegradation in the sub-surface attenuation of a landfill leachate plume has been demonstrated by Lyngkilde and Christensen (1992), and Williams *et al.* (1991) In the former study of a landfill in Vejen, Denmark, the majority of the dissolved organic carbon was degraded within the first 100m of the plume. In the latter study, although the TOC changed little throughout the plume, phenol degradation was apparent and coincided with the formation of benzoates. The data suggested that intermediates formed from the degradation of the original contaminants were further metabolised to aliphatic compounds. However, important though it undoubtedly is, the extent of microbial degradation of organic materials within aquifers will vary considerably according to local conditions and may be limited in unbuffered or poorly buffered strata where bacterial growth may be limited or prevented.

5.5 Leachate Control

The best way of controlling pollution of the environment by landfill leachate is through prevention, for as identified above, once leachate has entered groundwater it is both difficult and expensive to clean-up the groundwater; in most circumstances it is extremely unlikely that the polluted groundwater would be restored to its former state. Measures designed to prevent leachate migration should be an integral part of a site design, and the proposed European Landfill Directive (CEC, 1991) and sub-title D of RCRA, require that all landfill sites should be operated as containment sites, in which all liquids are held within the site. In recent years, most landfills in the "developed" world have been designed on this basis (but see "dry-tomb" - chapter 2). In these circumstances, unless moisture is very carefully controlled there will be a need to collect and treat leachate as it forms, if the site is not to run the risk of filling with leachate. This requirement has had the effect of stimulating interest in leachate treatment.

The first stage of effective leachate control is the design, engineering and construction of an effective leachate management system. A document prepared by the North West Waste Disposal Officers (1991) describes such a system in some detail.

The principles of landfill design have been described previously, while the relative advantages and disadvantages, and the future direction of landfill design is described in Chapter 7. However for most landfill designs, there will be a need to control liquid ingress, liquid collection, and liquid treatment. The use of landfill liners is an important element of each of the above processes.

5.5.1 Landfill Liners

To be effective in either limiting liquid ingress or egress and thus help to ensure effective leachate management, landfill liners/covers should be impermeable (or demonstrate very low permeability), durable, and resistant to chemical attack and puncture.

The liner systems may utilise natural materials such as compacted clay or shale, bitumen, soil sealants (bentonite) or synthetic membranes otherwise known as geomembranes, or flexible membrane liners (FML's). The choice of lining system will vary according to local conditions and national legislation. The European Landfill Directive requires that liners be used to prevent "pollution of the soil, groundwater or surface water and ensuring efficient collection of leachate". Specific requirements are that the base and sides of a landfill accepting non-hazardous waste (including household waste) should be lined with a composite liner (see below), where the mineral liner is at least 1m thick and has a maximum permeability of 10^{-9}m.s^{-1}. Under sub-title D of RCRA lining requirements are made to ensure that contaminant levels do not exceed those shown in Table 10, and require that a composite liner system is used that is designed to maintain less than 30cm of leachate over the liner. The requirements in the USA are, in general, more specific, and specify minimum thickness requirements for different types of geomembranes, as well as the specifications for the mineral liners (which are similar to those prescribed in the European Landfill Directive). Both the European and USA systems of control have a strong risk assessment element, allowing design criteria to be specified on a site-by-site basis, but the requirements under sub-title D of RCRA are much more prescriptive with respect to the methods and controls to be used during construction.

Natural Liners used in landfill construction are comprised mainly of materials such as clays (including the various types of bentonite) and shales. The advantages of these materials are that they have a low permeability (but one that is much higher than geomembranes), are relatively resistant to chemical attack, they are relatively puncture proof, and because of sorption properties, they have an inherent attenuation capacity that can be used to advantage in treating leachate as it migrates through the barrier. Recent studies (Peters, 1993) have shown that natural clay bedrocks are little affected by landfill leachate, that heavy metals are immobilised in the upper layers and that for those systems studied, that the migration of soluble compounds was retarded by a factor of between 20 and 60, and clearly demonstrates the attenuating capacity of mineral liners. Bentonites are also available as a "sandwich" between layers of various geotextiles, where it is claimed that their high swelling capacity makes them ideally suited as a component of a basal liner system. However, these sandwich-type materials are relatively thin (seldom >1cm), and may be easily punctured. Bentonite can also be used as an additive to existing soils to decrease permeability. However, as bentonite appears to be susceptible to deterioration when in contact with landfill leachate, resulting in increased permeability, these materials are likely to be of greater value when used as landfill capping material, where potential for puncture is much less, and where resistance to the effects of leachate on material integrity is not a concern.

Mineral liners are normally specified in terms of their **permeability**, and therefore, by definition, leachate will migrate through these liners. Thus mineral liners alone will not act as true containment barriers. However, long-term containment is not necessarily desirable (nor actually achievable), and the attenuation capacity of these liner materials will be an important element of a sustainable landfill, as defined in Chapter 2. Indeed, it could be argued that current trends towards pre-treatment of waste to reduce the associated hazards, or the **effective** management of a bioreactor landfill, would require little more than a mineral liner system to achieve sustainable (acceptable risk) landfill. This scenario would favour a return to dilute and attenuate principles. However, in any dilute and attenuate landfill there will be a proportion of the leachate that will migrate through the mineral liner before any effective attenuation has occurred. Although this may take a number of years, and although some attenuation may occur within the natural material, as described above, the associated risk to the environment may be unacceptable As a result, the use of geomembranes would appear to offer a number of advantages to the mineral liner alone - on the understanding that the function of the geomembrane is to control short-term risks, while waste and landfill management pratices for leachate control (including waste minimisation, recycling, pre-treatment, and bioreactor landfill) manage the long-term risks.

Geomembranes (typically high density polyethylene (HDPE), or medium density polyethylene (MDPE)) have low permeability to most materials, they are relatively easy to install and are relatively strong with good deformation characteristics. They suffer from the disadvantages of being slightly permeable to some solvents, they can be relatively easily punctured, they have little or no attenuation capacity, and their long term performance in landfill sites is not fully proven (this will not be a problem if such liners are used with the sole intention of providing short-term security). Impermeable/flawless seams between adjacent sheets of material can be difficult to achieve, especially in unfavourable weather conditions, and good construction quality assurance is a vital element of landfill construction, especially when using geomembrane liners. As well as problems with seaming, temperature-dependent shrinkage, expansion of the polyethylene, problems with tears and punctures during the lining process, and the relative ease with which this happens, requires that great care is taken during the lining process and that effective planning has been undertaken. These problems also render geomembranes unsuitable for use as a single liner in most circumstances. In the short-term risk control strategy, it will be essential to ensure that any damage is detected and repaired before waste is emplaced. The implications of liner damage for the escape of leachate are shown in Table 11. It can be seen from this table that for an "average case", calculated flow rates of 25,000 $l.ha^{-1}.day^{-1}$ for a geomembrane alone reduce to $47l.ha^{-1}.day^{-1}$ for a composite liner, thus demonstrating potential weaknesses of using a geomembrane alone and highlighting the benefit can be derived when using geomembranes in conjunction with natural liners.

Table 11.   Calculated flow rates through liners. Flow rates in $l.ha^{-1}.day^{-1}$

Type of Liner	Best Case	Average Case	Worst Case
Geomembrane Alone	2,500 (2 Holes/ha)	25,000 (20 Holes/ha)	75,000 (60 Holes/ha)
Compacted Soil Alone	115 ($k = 10^{-10}$/s)	1,150 ($k = 10^{-9}$m/s)	11,500 ($k = 10^{-8C}$m/s)
Composite	0.8 (2 Holes/ha, $k = 10^{-10}$m/s. Poor Contact)	47 (20 Holes/ha, $k = 10^{-9}$m/s. Poor Contact)	770 (60 Holes/ha, $k = 10^{-8}$m/s, Poor Contact)

After Street (1993)

Liner combinations can often be used to enhance the effectiveness of the overall containment system. The reduction in flow rates that can be achieved by using a composite liner, rather than a geomembrane alone or mineral liner alone, is shown in Table 11. The choice of liner combination will be dependent upon local conditions and policy. Mineral liners and geomembranes are often used together for this reason and represent a popular choice in the construction of new landfill sites, and are a minimum requirement of both USA and European legislation. The design of landfill liner required in the USA for unapproved states is shown in Figure 10. By using the geomembranes in conjunction with mineral liners, many of the disadvantages of each can be overcome and an effective barrier can be achieved. Such composite systems utilise the very low permeability of the geomembrane, and the attenuation properties of the mineral liner, and minimise any harmful effects that would arise as a result of damage to the geomembrane. However, as described in Chapter 2, it is important to recognise the anticipated effective lifetime of the geomembrane at the design stage so that the landfill may be managed in such a way that when liner failure occurs, the associated risk to the environment is acceptable.

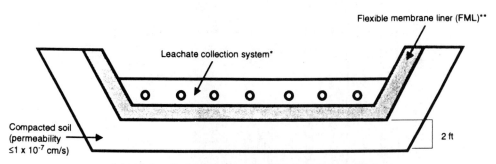

    * Leachate collection system must maintain leachate level <30 cm.
   ** FML must be at least 30 mil thick; FML consisting of high density polythylene (HDPE) must be at least 60 mil thick (mil = 1,000[th] of an inch).

Figure 10.   Composite liner and leachate collection system design in unapproved states (USEPA, 1994)

Reproduced with permission of USEPA

Multiple liner systems such as that shown in Figure 11 have also been used on many landfill sites. In many of these systems, a leakage detection layer is placed within the liner system; proponents argue that this facilitates detection, and/or collection of leachate upon failure of the upper liner layer. However, this

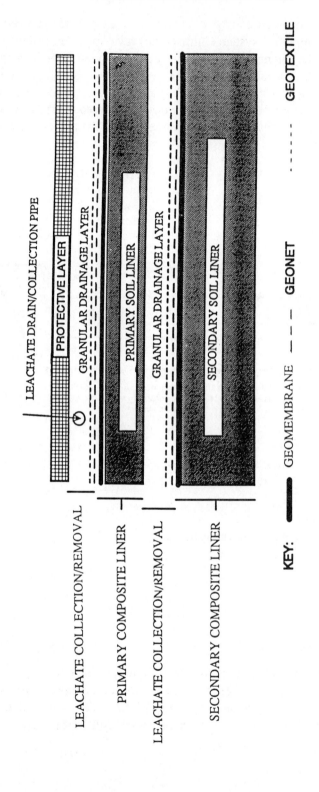

Figure 11.   Multiple Landfill Liner System

argument fails to recognise that a single hole within the upper liner will allow a head of leachate to accumulate across the base of the site, as the detection layer "floods". The associated risk to the environment, when leachate has the potential to leak across the entire base of the site is much greater than that associated with a similar failure in the much simpler composite liner system which would result in leakage across the 1cm hole, and not across what may be many hectares. For this reason, for the relative ease of construction, for the reduced potential for slippage of layers (and related catatrophic liner failure), and for cost-effectiveness, the composite liner appears to offer the most appropriate liner system for most landfill situations.

Another feature of the multiple liner system shown in Figure 11 is the use of various geotextiles. These geotextiles are often used to protect membrane liners, and other damage-sensitive elements of the liner. When used as protection in this way, they can offer a number of advantages. However, they have also been used as filters to restrict the movement of fines into the collection system. In this case, the geotextiles may become blocked themselves (by both fines, and biomass that may grow in the leachate environment), and may thus severely restrict leachate movement, and in this way worsen the situation they were intended to resolve. More detailed information on the uses, advantages and disadvantages of liner systems can be found elsewhere (North West Waste Disposal Officers, 1986; DoE, 1995b; USEPA, 1994; Seymour, 1992).

Having effectively sealed a landfill site against leachate escape (at least in the short-term), and anticipating the need for leachate collection and treatment, the installation of an effective leachate drainage system is essential. In many cases, leachate drainage systems have been designed on a herring-bone pattern (or similar), but the use of a drainage blanket across the whole base of the site is now considered as best practice in many countries. Collection drains can be placed within the blanket, and leachate can be conveyed to a central sump for pumping off-site. For effective leachate collection, preparation of all site bases should ensure that gradients are generally not less than 2%, and the material used for the construction of the drainage blanket should be relatively fine-free, and of such a size as to show good conductive properties. This will help to ensure that collection sytems are not blocked.

A more detailed consideration of drainage blankets has been undertaken by North West Waste Disposal Officers (1991).

5.6   Leachate Treatment

With reference to the source-pathway-receptor model described briefly in Chapter 2, the previous section on control of leachate migration has dealt with the "pathway" element, and more specifically how the leachate may be prevented from entering the pathway (the environment surrounding the landfill). From the preceding discussion, it is clear that prevention of migration cannot be guaranteed, and that once leachate has escaped to the surrounding environment, control and effective remediation is seldom achievable. Furthermore, as we cannot control the sensitivity of the receptor (humans, plants etc.), and because removal of the receptor from a polluted area, or other mechanisms to protect the receptor are both difficult and expensive, the "source" element is the remaining element of the model that can be addressed. Therefore for effective management of the overall environmental risk posed by landfill leachate, leachate removal and/or treatment is essential. For reasons mentioned previously, the treatment of landfill leachate is likely to become a more common process at landfill sites in the future. Currently, the leachate from many landfill sites that do not operate on a dilute and attenuate basis is discharged directly to foul sewer, and for which a charge is made. In this case, the leachate removal is to an off-site treatment facility (the sewage works) where treatment to meet agreed environmental limits can be achieved. For the remainder of the leachate, recirculation, chemical treatment and biological treatment can be undertaken, with eventual discharge to sewer or surface water course, according to local conditions, controls, and the efficiency of the leachate treatment process.

5.6.1 Leachate Recirculation

One of the simplest forms of leachate treatment is that achieved by leachate recirculation. Because the landfill acts as an anaerobic biological reactor, recirculation of leachate within the waste facilitates further biological degradation and physico-chemical treatment of the remaining undegraded organic and inorganic leachate components. This reduces the hazardous nature of the leachate, and through the redistribution moisture increases the potential for waste to become wet. It is likely that uneven moisture distribution throughout the waste body is one of the major reasons for the long time-scale of landfill gas production, as those regions that are wet will produce landfill gas, while those that are relatively dry will produce relatively little gas. Enhanced moisture distribution through recirculation will reduce this heterogenous characteristic and will also act to effectively increase mixing within the waste.

Enhanced waste degradation and the effects of recirculation has been described in Chapter 3, where studies suggested (Klink and Ham, 1982) that moisture **distribution** is a key factor in ensuring effective waste degradation through optimising conditions for microbial activity. Therefore by ensuring moisture distribution throughout the waste, more effective treatment of the solid waste component may be anticipated. Although the waste, when emplaced, will have a moisture content below field capacity, leachate will develop in the base of the site as the moisture is squeezed out of the lower layers of waste. In a containment site this leachate can be collected, pumped to the surface, treated, and redistributed throughout the waste via a system of under-cap pipes and drains.

If spray irrigation is used as a means of recirculation, then volume reduction may also be achieved, as a result of evaporation. However, the potential for atmospheric pollution by volatile organic compounds should not be overlooked. In some cases, surface sealing as a result of the formation of a crust of precipitated metal salts can be unsightly and can create management problems. The distribution of leachate through a distribution system placed below the cap may be more difficult to engineer, but because of the above, is in most cases preferable to surface irrigation methods. The largest problem associated with the use of a sub-surface recirculation system, is likely to be associated with settlement, and maintenance of the integrity of the pipework.

5.6.2 Chemical Treatment

Chemical treatment of landfill leachate is not widely practised and has often been used in the past to deal with specific treatment requirements such as the removal of odour. Chemical treatment is normally undertaken using either precipitants and flocculants or chemical oxidants. The former are used primarily for the removal of dissolved metals (and associated colour) and include precipitants such as calcium and sodium hydroxide. Chemical oxidation can be used for the removal of odourous compounds such as sulphides but is not used frequently. More recently, hydrogen peroxide has been used as an additive to leachate before discharge to foul sewer in order to reduce problems associated with methane gas as it moves out of solution, and the associated explosion risks. This requirement has been placed upon some landfill operators, although it seems that little work has been done to assess whether the risk associated with this process is significant, and whether or not the addition of hydrogen peroxide is effective (or the most cost-effective) means of dealing with the presumed problem.

Many of the treatments effected by chemical addition may also be achieved by biological treatment, and biological treatment is currently favoured by most landfill operators as the most appropriate means of reducing the leachate-associated hazard.

5.6.3 Biological Treatment

The treatment of leachate via recirculation (above) is an example of an anaerobic biological treatment process, and in recent years the use of biological processes for leachate treatment has increased considerably. In the recirculation process, and before addition of the leachate to the landfill, it is often pre-treated by an aerobic biological process which complements the anaerobic activity within the waste, and results in more effective leachate treatment. The effectiveness of biological processes will vary according to the nature of the leachate, environmental conditions such as temperature, and the effectiveness of the design and control of the reactor, but leachate treatment to standards suitable for discharge to surface water-course have been achieved. Biological treatment to remove BOD, ammonia and suspended solids (with or without recirculation) is a common form of leachate treatment and aerobic processes tend to predominate. Aerated lagoons and activated sludge processes are often used for BOD and ammonia removal, while under conditions of low COD, rotating biological contactors (RBC's) have been especially effective in the removal of ammonia. The removal of recalcitrant organics and trace hazardous organics not treated by biological processes can be effected by chemical treatment, activated carbon and reverse osmosis. In theory, anaerobic leachate treatment processes offer some advantages over aerobic processes in that the energy costs are lower and the amount of biomass produced is also low. However, aerobic processes are often easier to control and operate, and this tends to favour the use of these processes, despite the **relatively** high electricity costs. Also, because the landfill is in itself an anaerobic biological reactor, little additional benefit could be anticipated from the use of a further anaerobic treatment process.

Three of the most commonly used biological processes are the aerated lagoon, the rotating biological contactor and the vegetated ditch (or reed bed).

The aerated lagoon is a simple and robust system and can achieve effective removal of organic material (BOD, COD) and ammonia (Robinson *et al*, 1992), although they do have relatively high energy requirements. In most cases careful control of the BOD should be exercised in order to ensure good sludge settlement and effective control over the process. The hydraulic retention time varies in most cases between 10 and 20 days and the process can be susceptible to seasonal temperature changes. Removal levels are generally good and can average 99% for BOD, 97% for COD and 93% for ammonia (Ashbee and Fletcher, 1993).

Table 12 shows figures relating to a specific leachate and treatment plant, and demonstrates the potential effectiveness of these systems.

RBC's have proved to be very efficient for the nitrification of ammonia but are unsuitable for ammonia treatment at high BOD loadings, and thus relatively high C:N ratios. In such cases pre-treatment may be used to reduce the carbon content of the leachate. Hydraulic retention times are relatively low and the energy requirement is also relatively low, thus conferring distinct advantages over the aerated lagoon processes. However they tend to be more mechanically complex than aerated lagoons and therefore more prone to mechanical failure.

The vegetated ditch, otherwise known as a soil-plant system or reed bed have been used for leachate "polishing" but have limited use as a primary treatment process. The treatment relies on the ability of reeds (often Phragmites australis) to stimulate bacterial growth in the root zone by the transfer of oxygen to the roots. The bacteria are the effective treatment organisms. These systems are now finding greater application in the treatment of landfill leachate, but are not currently used widely.

Table 12   Compton Bassett leachate treatment plant: Mean composition of influent leachates and of final effluent, March 1987 to December 1988 inclusive (flow-weighted mean values in mg/l, except pH value)

Determinand	Cell 3	Cell 4	Cell 5	Effluent
pH	7.4	7.4	7.2	8.3
SS	123	180	430	38
VSS*	61	105	356	23
COD	746	1310	4298	268
BOD	133	313	2346	10.4
Ammoniacal-N	565	809	865	3.4
Kjeldahl-N	633	895	935	8.1
Nitrate-N	0.9	1.5	0.3	177.5
Nitrite-N	0.04	0.07	0.16	0.46
Calcium	101	131	314	98
Iron	16.2	16.9	71	0.5
Zinc	0.9	0.8	3.8	0.3

* Volatile suspended solids

With Permission. Modified from Robinson *et al* (1993)

      The use of anaerobic biological systems is likely to be limited to denitrification processes, as ammonia removal is poor, there being no recognised metabolic pathway in anaerobic systems. Because of the above, the removal of ammonia from landfill leachate is difficult to achieve. Data in Table 12 taken from a surface-aerated leachate treatment plant at a landfill in the UK, show that aerobic processes can be successful in some circumstances, and that ammonia levels can be reduced to the point where they are suitable for discharge to foul sewer, or in some some cases to surface water courses. However recent studies have identified a potential alternative (Knox and Gronow, 1993). The principle of the treatment process is that ammonia within the leachate is treated in an aerobic process to convert the ammonia to nitrate (nitrification). The nitrified leachate is then recirculated through the anaerobic waste to convert the nitrate to nitrogen gas which is inert and can be removed safely and easily from the landfill environment. This is represented diagrammatically in Figure 12. In pilot-scale studies using fresh refuse, Knox and Gronow (1993) showed that complete denitrification at rates of at least $12.5gTON.m^{-3}.d^{-1}$, could be achieved, even at low temperatures (10^0C), while ammonia was simultaneously flushed from the waste. In fully decomposed refuse there was little capacity for denitrification but removal of COD and colour was significant.

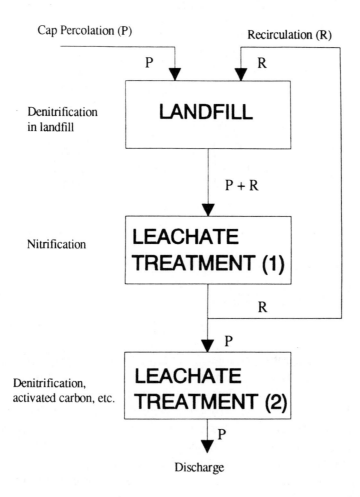

Figure 12   High rate leachate recirculation concept (Knox and Gronow, 1993)

Reproduced with permission

CHAPTER 6

OTHER LANDFILL HAZARDS

6.1   Introduction

In any discussion of the environmental problems associated with waste disposal to landfill, the major potential sources of pollution are landfill gas and landfill leachate although problems associated with litter, birds (especially gulls), vermin, noise, dust, traffic movements and visual impact (especially when land-raising) could all have a detrimental effect on quality of life, and could constitute a statutory nuisance to those living and working in the vicinity of a landfill. Control of these hazards is often relatively easily achieved through the operation of appropriate landfill management techniques, which will be discussed in this chapter. A crucial difference between landfill gas and leachate, and the other landfill-associated environmental problems is that problems associated with the latter will be confined, to a very large extent, to the operational phase of the site, whereas those associated with gas and leachate may be present throughout the lifetime of the site and beyond.

6.2   Litter

Poor litter control can result in widespread contamination of the surrounding area in a way that can be offensive to all those in the neighbourhood. Problems associated with windblown litter can be minimised by the use of good operational procedures, such as the use of daily soil cover, and litter screens, while the effective sheeting of vehicles will limit problems associated with litter on the highways. Various proprietary cover materials are now available that may be used as alternatives to soil cover, and include paper-pulp preparations that may be sprayed onto the waste as a slurry, and which will, in theory, degrade when covered so that the potential for perched water tables, and preferential gas migration pathways are minimised. However, materials such as this are more expensive than locally sourced cover material and so far are relatively unproven.

6.3   Dust

Problems associated with dust are largely restricted to periods of dry weather, and are mainly confined to the area of the landfill. Dust can be suppressed by the use of wheel washers, the application of water to dry and dusty roads as and when required, appropriate handling procedures for dusty waste. The effectiveness of these measures will be dependant to a large extent upon the diligence with which they are undertaken, and the prevailing environmental conditions.

6.4   Birds, vermin and insects

Problems associated with birds, vermin and insects are related to the type of waste being filled and will be greater when the waste contains a significant proportion of organic material. Problems associated with vermin and insects can be minimised by the effective use of daily cover, and regular visits of the site and surrounds by pest-control officers. In many cases in the developed world, problems associated with the above have decreased as a result of more rapid filling of a specific area, as result of the cellular construction of landfills, and the reduced potential for these pests to complete a life-cycle before being buried under more waste. If this argument is accepted then the main value in the use of daily cover is in the minimisation of windblown litter, and in the light of the void space that daily cover occupies, and the potential problems it creates (e.g. relatively impermeable layers within the waste), it may be worthwhile considering alternative ways of dealing with this problem.

While the hazards associated with vermin and insects are essentially related to the spread of disease (such as Weil's disease), those associated with gulls are also concerned with hazards to aircraft at those landfill sites situated near airport runways. In either situation resolution of the problem is not easy. At some landfill sites, nets have been erected over the working face as well as around the periphery, in order to deny the birds access to the working face. This method appears to have been successful but does place restrictions on the effectiveness of operation of the site and may not be a feasible option for all sites. Birds of prey have been used with some success, but in most cases the target birds return once the bird of prey (the use of which can be expensive) is removed. Other methods of control such as the use of automatic bird scarers, bird distress calls and the use of shotguns have all met with little or no success and gulls at landfill sites remain a problem with potential health implications for those living within the vicinity.

6.5 Noise

The noise at landfill sites is essentially associated with vehicle movements, either as a result of the throughput of vehicles, or as a result of reversing bleepers on plant. Such noise can be most easily controlled by the effective use of natural or artificial barriers such as soil embankments, tree stands and good use of the local topography, although there will be obvoius problems associated with landraising schemes, where there will be potential for noise to travel further than with normal landfill schemes, and where the construction of barriers to sound travel will be more difficult.. Controls may also be placed on the operational hours of the site.

6.6 Visual impairment and odour

Problems associated with the visual impairment of the area around a landfill site are best resolved by planning and design measures whereby the emplacement activities are screened from general view by means of natural or artificial screens. This will seldom be possible when land raising and should thus be an important consideration at the planning approval stage. For large landfill sites producing large volumes of gas, it will not be possible to easily mask the large stack that may be 5 or 10 metres high and this will also require consideration at the planning stage. At the other end of the scale, landfill sites that do not utilise a flaring system may experience problems with odour from emissions from vents within the waste. In these circumstances the installation of a more effective gas control system which would utilise a gas flare, would be required. Under the proposed European Landfill Directive passive venting of landfill gas will not be allowed unless an environmental assessment shows this to be acceptable, while in order to control fugitive gas emissions from landfill sites within the USA, stringent controls on surface emissions from landfill sites have recently been introduced by the USEPA (1990b). Within the state of California, landfill operators are now required to monitor the trace component composition of landfill gas and to assess the impacts on air quality. This serves as an example of the increasingly stringent controls being placed on landfill operations, both during and after the active life of the site, in order to minimise the environmental impacts.

Other sources of odour at landfills include leachate and leachate treatment systems, sewage sludges, and waste materials, especially those that have, for whatever reason, decomposed prior to landfilling.

For each of the above factors, the associated risk will be dependent upon the sensitivity and location of the 'receptor', which may include green fields, wildlife and specific flora and fauna, buildings, ground and surface waters and local populations. Factors such as the prevailing wind direction, distance to the nearest sensitive target and natural topographical and hydrogeological features will therefore be important considerations at the planning stage, and if undertaken effectively should significantly reduce many of the environmental problems discussed above. This will also be true of gas and leachate related problems.

CHAPTER 7

SUSTAINABLE LANDFILL - THE WAY FORWARD

7.1   Introduction

The previous six chapters have explored factors that may impact upon the design and operation of a sustainable landfill, which was defined in Chapter 2 as one in which "the associated risks have been measured and determined to be acceptable". According to this definition the sustainable landfill may vary from country to country and region to region, according to a range of factors including the nature of the waste to be landfilled and local geology/hydrogeology. Also, according to the risk model described in Chapter 2, and as elaborated in Chapters 4 and 5, one of the areas with greatest potential for risk reduction is control of migration to the surrounding environment. The choice between prevention or control is dependent upon the extent to which containment of waste and waste products can be successfully achieved.

7.2   Containment or dispersal?

The principles of landfill practice were discussed in Chapter 2, and amongst the differing views on landfill best practice there seems to be at least one issue where there is a consensus of opinion. That is, whether for monofill or co-disposal, bioreactor or dry tomb, the use of a containment system is essential in the short term. However, according to local conditions and how "containment" is defined, a natural clay barrier **may** provide sufficient containment to satisfy the requirements of sustainable landfill. As discussed in Chapter 2, the dilute and attenuate principle (as an immediate management option rather than a long-term inevitability) can only be justified where a site-specific risk assessment has shown this to be feasible. Such a landfill may be appropriate for well-defined inert waste, where the expense associated with landfill lining could not be justified. This again emphasises that for sustainable development to be achieved there must be **development,** and that the cost of environmental protection must be balanced against the associated risks. Where dilute and attenuate is not feasible then the use of a containment system provides time in which to manage the waste to reduce the risk and, for the foreseeable future, will be an integral part of the majority of sustainable landfills.

It is important to recognise that while many landfills are currently designed as containment systems, the containment barriers will **ultimately** fail. Upon failure, any liquids within the site will migrate into the surrounding environment with a consequent risk of environmental pollution. It is therefore important to design and manage landfills in such a way that upon liner failure, the consequent risk is acceptable. Data in Table 13 show that for a mineral liner of hydraulic conductivity = 10^{-9} m. sec^{-1} the flow through time for a 2 metre thickness is 14.8 years. This emphasises the advantage of using composite liners, where their use will increase the effective containment period, restrict liquid flow rates once the containment system fails, and recognising the flow-through time (Table 13), and the sorptive properties of the natural liner (Chapter 5), will attenuate any leachate as it escapes.

Table 13.   **Effects of liner thickness on flow-through time**

Thickness (m)	1	2	5	7	10	15
Hydraulic gradient	2.0	1.5	1.2	1.14	1.10	1.07
Leakage rate (mm yr^{-1})	63.0	47.3	37.8	35.9	34.7	33.7
Flow-through time (years)	5.5	14.8	46.0	68.0	101.0	156.0

It is also important to recognise that in accepting the 'ultimate liner failure scenario', there is an implicit acceptance that the sustainable landfill is one that **ultimately** relies on the dilute and attenuate principle, and that the prime objective of a truly sustainable landfill is the reassimilation of the waste into the surrounding environment. This concept has been discussed previously by Campbell (1992). Recognising this, it would seem that if we are able to more effectively stabilise waste within the landfill, or to pre-treat waste prior to landfill, the simple composite liner has many advantages over more complex multi-layer systems, especially those with back-up leachate collection systems (weaknesses of these systems have been discussed in Chapter 5). It remains to be seen whether or not the objective of reassimilation into the environment is achievable, but in the meantime, it reminds us of the size and nature of the task ahead, and provides a goal which may direct intermediate stages in landfill development.

7.3   Dry-tomb landfill, bioreactor or pre-treatment

If containment of waste is one area where risk reduction can be realistically achieved, a second area with greater potential is that of hazard reduction. This can be achieved through careful control of waste inputs and through appropriate choice of landfill design and operation (e.g. bioreactor landfill, pre-treatment). The choice of design and operation of a sustainable landfill will be affected by the nature of the emplaced waste. If truly inert waste is deposited there will not be a hazard and sustainability will be easily achieved. Waste minimisation and recycling will impact upon the composition and hazardous nature of waste and the effects of these practices will affect the choice of sustainable landfill design.

In the following, the discussion is confined to the disposal of MSW (as defined in Chapter 3), while issues associated with hazardous waste are addressed later.

In theory, management of a landfill in such a way that the risks are acceptable can be achieved in a number of ways, and there are currently three schools of thought on the sustainable landfill; One that argues for treatment of the waste within the landfill (bioreactor landfill), in which case the landfill is used as a treatment facility rather than a disposal facility; a second that argues for waste pre-treatment and the emplacement in landfill of the inert residues from the treatment process; in this case the landfill acts as a disposal facility; and a third that argues for dry waste storage, in which water is excluded ("dry-tomb" - Chapter 2). Unfortunately, for many the sustainable landfill is defined according to one of these three concepts. This limited view risks the establishment of entrenched camps of supporters for a particular method, and fails to recognise that other factors may be considered. A sustainable design and operation in one region may not be sustainable in another! If this argument is accepted then it follows that a single set of guidelines (e.g. for Europe) is not reasonable - by virtue of different waste management policies and strategies, different waste arisings, different geology etc., and may well hinder appropriate and cost-effective sustainable landfill development.

Under the first of the above scenarios, the advantages of using a landfill as the treatment process include:

- single stage treatment and disposal
- potentially lower costs
- potential for energy generation from landfill gas
- potential to restore blighted countryside to effective use
- potential to effect treatment of codisposed industrial waste

while disadvantages include:

- significant pollution potential from landfill gas and leachate
- potential for nuisance from vermin, birds, odour etc.
- greater explosion and fire hazard
- difficulties in achieving effective destruction of waste materials
- potential for extremely long-term liabilities

Under the second scenario (and assuming that incineration is the primary pre-treatment process), the advantages include:

- volume and mass reduction, and therefore, more efficient use of available void spaces
- potential for heat and enegy generation and utilisation
- minimisation of gas-related risks upon final disposal

Disadvantages include:

- production of secondary waste streams (e.g. alkaline waste from scrubbers)
- higher costs than landfill (at the time of writing)
- the need for transport to final disposal (with associated environment and cost implications)
- strict emissions control required
- difficulties in achieving 100% destruction of waste materials
- high chloride and metal content of waste ash, with associated long-term liabilities

The dry-tomb scenario is discussed separately below.

The lists above are far from exhaustive, but illustrate the problems involved in selecting the best option - the BPEO! Close examination and comparison of any one of the above factors is difficult and time-consuming. For example, in its 17th report, the RCEP (1993) compared the emissions of greenhouse gases from landfill and incineration and showed that the "greenhouse impact" from incineration was far less than that from landfill (Figure 2). However in reaching this conclusion several bold assumptions were made, and by changing some of these assumptions, Wallis (1994) showed that the reverse result could be obtained (Figure 13). If statistics are right .... and statistics prove that they are!! The problem here is that much of the quantified information required for effective assessment of the options are lacking. For this reason, and unless a great deal of time and effort is spent on generating the necessary data, determination of the most sustainable waste management option will always be difficult.

7.3.1 The "dry-tomb landfill"

The dry-tomb landfill described in Chapter 2 is an extreme approach to the control of landfill pollution and serves as a useful example of landfill management that is not sustainable. This approach is not sustainable, because it effectively constitutes waste storage rather than treatment, and will ultimately result in pollution of the environment when environmental conditions change and the containment system fails. This option seems to be an appropriate and (by virtue of lower requirements for leachate and gas control) relatively cheap option in areas of low rainfall, but fails to recognise that at some time in the future, both local and global climate changes may result in significant leachate production and environmental pollution. However, in the USA where this approach has been adopted under RCRA, recent evidence suggests that both the scientific community (e.g. Maurer, 1993), and the relevant federal bodies within the USA, are reconsidering the "dry-tomb" approach to landfill. The inadequacies of the dry-tomb approach have been highlighted by Lee and Jones-Lee (1993), and include:

- landfill cover cannot be expected to restrict water infiltration to the extent that leachate does not form
- landfill liners will not prohibit migration of leachate into the surrounding environment
- the key components of a "dry-tomb" landfill system are not available to inspection
- once contaminated by leachate, groundwater cannot be effectively cleaned up
- MSW in a "dry-tomb" landfill will be a threat to groundwater quality forever
- MSW leachate can pollute very large amounts of groundwater

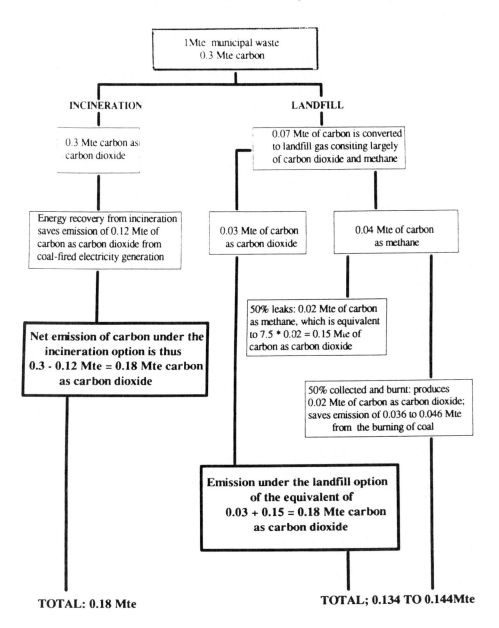

Figure 13.   Incineration versus landfill: Emissions of greenhouse gases
              (Wallis, 1994)

Although the USEPA also acknowledge the limitations of containment systems and weaknesses in the dry-tomb approach, the requirement for the "dry-tomb" approach to landfill management remains.

An alternative dry landfill technique has been suggested by Stegmann (1993). In Germany, only that waste remaining after "extensive waste avoidance programmes" will be landfilled, and only then after primary treatment (Stegmann, 1993). Thermal pre-treatment is the preferred process, but while the infrastructure and facilities for incineration are developing, biological pre-treatment is being discussed as an alternative option (Stegmann, 1993). The system proposed by Stegmann (1993) utilises aerobic and anaerobic biological processes - according to the nature of the waste, after which the pre-treated waste is disposed to highly engineered landfills with mobile rooves to keep out precipitation.

According to Stegmann (1993) primary biological treatment has the following advantages:

- Improved compaction densities
- low settlement
- low gas production rates
- low concentrations of biologically degradable organic compounds in the leachate
- improved conditions for landfilling through reduction of nuisance factors such as dust, odour, etc.
- lower volume of residual waste for landfill

It is accepted that some organic matter will remain. Stegmann (1991)(cited in Stegmann, 1993) estimated that 10 % of degradable material would remain with a potential to produce 15-20 $m^3.t^{-1}$ of gas, and that gas and leachate monitoring and control systems will be required. Although the risk associated with this type of system is less than that with current conventional landfills, there is an unswerving faith in the durability of the containment system and the stability of institutional control, and an apparent acceptance that long-term monitoring is acceptable. This system does not allow for eventual dispersion of the stabilised waste material into the environment, and according to the definition adopted here, does not represent sustainable landfill development.

The "dry-tomb" approach, which merely transfers the responsibilities and liabilities of waste management to future generations does not represent an appropriate way forward for landfill management.

7.3.2   The Bioreactor landfill

In the system proposed by Stegmann (above), biological degradation is seen as an appropriate means of pre-treatment, and in recent years, the concept of the bioreactor landfill which utilises controlled biological degradation within the landfill site has gained increasing support in some circles, as a means of **treating** waste. It is difficult to pinpoint from where the concept arose, and it seems to be an idea that has been developed and modified within the waste scientific community over a period of time. The underlying principle is that by optimising conditions (discussed in Chapter 3, but especially moisture content) within the waste, more rapid and complete degradation of waste may be achieved. The general objective is to produce a "stable waste" within 30 years of emplacement (See Chapter 2), and thus ensure that the risk to the environment will be at an acceptable level when liner failure occurs. However, the figure of 30 years has been derived variously in different countries, and there is no single good reason why this figure should be the accepted goal. The only sensible requirement is that when the containment system breaks down, the associated risk to the environment should be acceptable, and thus the durability of the liner system will be a key factor. As discussed above, if a composite liner is used (Chapter 5), the effective lifetime of the contaiment system will be much greater than that of the geomembrane component, and the 30 year timescale for stabilisation could be greater.

Whether landfills are designed for stabilisation within 30 years, 100 years, or 500 years, the implications of failing to more effectively manage landfill waste have been spelt out by a number of workers (Harris *et al*, 1994; Lee and Jones-Lee, 1993). For the bioreactor landfill a number of key factors will be critical to its success. According to Walker (1993), probably the most difficult component of leachate to eliminate is ammonia; for a 10m deep landfill, with waste of 40% moisture content, and an infiltration rate of 50mm per annum Walker (1993) calculates a hydraulic retention time of 80 years. Further calculations assume a typical ammonia content of 1000mg per litre and a target concentration of 1mg per litre, and use a washout equation quoted by Knox (1990b) to show that the time taken to achieve an acceptable 1mg per litre final concentration would be 552 years. Although it is necessary to make a number of assumptions in calculating the above, those made do not seem to be unreasonable, and one could anticipate that similar timescales would be required, for example, for landfill gas evolution to reduce to acceptable levels. The implications of this for landfill management are very significant in that they will require that both gas and leachate are effectively managed for hundreds of years after site closure - at considerable cost, and will further require that appropriate financial cover is maintained, which under the circumstances will be difficult to obtain, expensive, or both. If problems associated with the long time required for effective treatment of ammonia, carbon and other components of waste are to be resolved, the bioreactor landfill must become a more efficient and effective reactor than current landfills. Because current landfills are very poorly managed as bioreactors improvements should not be difficult to achieve. The uncertainty lies in whether sufficiently great improvements can be achieved to reach acceptable leachate contaminant levels. In theory (Chapter 3), the addition of water to a landfill will have two major effects: it will flush contaminants from the waste, and will create more favourable conditions for bacterial growth and hence waste degradation. In looking at the first of these effects, and using the assumptions relating to ammonia described above, Knox (1990b) showed that between 5 and 7 bed-volumes of water will have to pass through most landfills receiving degradable waste in order to reduce ammoniacal nitrogen levels to dischargeable concentrations. According to Harris *et al* (1994), if this is to be achieved within 30 years then "the mean hydraulic retention time must be less than 5 years, and for it to be effective, the passage of water must be reasonably uniform, reaching the whole of the waste mass eliminating zones of dead volume". This will require irrigation rates equivalent to 2000mm per annum or more, which is much greater than the effective rainfall in many areas where landfills are found. However, if leachate treatment and recirculation are employed then the equivalent of the required bed-volume changes can be achieved without the need for further water addition. The concept of enhancing biological degradation to reduce the organic content, and employing liquid flushing to remove degradation products and other compounds, including inorganics, is known as the **"flushing bioreactor"**. One of the first attempts to examine a bioreactor concept was undertaken by Bratley and Khan (1989). The principle of their "Landfill 2000" experiment is that after emplacement in shallow (2m) cells, the degradation of unsegregated waste would be controlled through temperature control and leachate recycle to produce a rapidly stabilised waste that could be removed and the cell used again for more waste. The degraded waste would be screened and sorted to remove recyclables and the remaining residue could be used to upgrade derelict land or or to infill mineral workings.

Since this work, the concept has been further developed, and a report has been produced by ETSU (1993) on a feasibility study to assess the potential of developing a bioreactor cell rotational landfill, referred to in the report as a "sustainable landfill". The feasibility study is based on highly engineered containment cells, where process control is effected through leachate recirculation, and where leachate from a methanogenic cell can be used to inoculate fresh waste. The waste stream is made up of the separated organic fraction of MSW, and other suitable waste such as gardens waste. The study concludes that "the sustainable landfill is considered to be a viable system for removing a considerable volume of material from the waste stream, and in so doing providing a useful energy source. It is best operated in the presence or in the vicinity of a materials recycling facility (MRF) and is complementary to current waste disposal options, particularly conventional landfilling and incineration". Thus in this case, the landfill evisaged in the study is not intended to replace the conventional landfill but to complement it. Unfortunately, this does not help to resolve the problem of the conventional landfill, and whether to operate it as a reactor, or a repository for pre-treated waste.

An alternative bioreactor concept - the fermentation/leaching wet cell (F/L wet-cell) has been developed by Lee and Jones-Lee (1993). This concept also aims to effectively stabilise the waste and leach the soluble potentially polluting components. The design requires that the waste is shredded prior to emplacement to try to ensure contact of the liquid with all waste components. According to Lee and Jones-Lee (1993) who cite Ham (1975), this will also eliminate the need for daily cover, and has the potential to increase the capacity of landfill by about 20%. The other key feature of F/L wet-cell is the use of a clean water system beneath the clay of the composite liner to maintain movement of water up through the clay, thus preventing any leachate leakages to escape from the containment system.

One of the major problems with this and other concepts of using the landfill as a treatment process is that while the theory is apparently sound, and may, in some cases have been demonstrated at laboratory-scale, it has not been proven in practice. All the evidence that is currently available suggests that to achieve a stable, relatively non-polluting waste in an acceptable period of time would be extremely difficult. However, in some large-scale studies, moderately encouraging results have been obtained; in work conducted at the Brogborough test cells in the UK (ETSU, 1993), it has been shown (Knox, 1995) that gas production in one of the waste cells doubled after air injection. At the same time, the methane content of the gas increased from approximately 50% before air injection to 55-60%. In the same study, the injection of water into another cell also increased the gas production rate. There are many potential, but no proven explanations for these findings. The important point is that the waste reactions within the cell can be manipulated to increase gas production, and hence waste stabilisation. To what extent we may be able to control the rate and extent of waste degradation in full-scale landfills still remains to be seen, and should form the focus of future research activities.

In Chapter 3, the landfill was described as a fixed-film bioreactor, and this is where the major problem lies - the landfill is an extremely poorly mixed system that relies on leachate recirculation (where practiced) to effect what mixing that may be possible. Under these conditions, waste degradation will proceed very slowly, even under conditions of optimum temperature and moisture (for which there are almost as many optima as there are investigations). There may also be problems associated with the waste composition, which although recognised, have not been tested at a reasonable scale. For example, in many experiments in which rapid waste stabilisation has been attempted, the work has been conducted on sorted waste in which inert material has been removed, yet to remove these inert structural materials from a full-scale landfill risks creating conditions in which the organic waste will compact to form a relatively impermeable layer that will prevent effective mixing within the site.

The F/L wet-cell of Lee and Jones-Lee (1993) requires that shredded waste be emplaced in order to improve fluid hydraulics, but until practices such as this are attempted at full-scale, it is difficult to predict what impact this may have on waste compaction, and in turn what effect this may have on fluid hydraulics. For example, if shredded waste is wetted before compaction, it is conceivable that a very low permeability waste layer could be created, and liquid flow would be severely restricted. Alternatively, if much greater rates of waste degradation are achieved, much greater rates and extent of waste settlement could be anticipated. This will have important ramifications for the design of a leachate distribution system that would have to cope with such settlement and remain intact. Landfill caps may also require re-design or regular mainteneance to deal with settlement that may exceed 50%. These are just some of the obvious potential ramifications of a change in one parameter in landfill management. Other factors that will have to be considered include the potential effects of increased waste temperatures on liner stability and gas collection systems, and potential changes in leachate composition and the knock-on effects for leachate treatment.

Thus it seems that while the theory is fine, and indeed may work in laboratory scale studies, it will be very difficult to achieve in practice. Even in studies such as the Brogborough test-cell investigations (ETSU, 1993), where some successful manipulation of waste reactions has been achieved, it may be many years before a stable waste is produced, if indeed this ever occurs. However, while there are many uncertainties, there are sufficient encouraging results to warrant further investigation, if not at this stage, whole-hearted support of the bioreactor concept.

However, irrespective, of the specific design, there are certain key design features that it appears, must apply to all bioreactor/wet-cell systems:

• effective containment is essential for as long as the waste presents an unacceptable risk to the environment
• good moisture distribution and control is essential

7.3.3 Pre-treatment

In most circumstances, incineration is likely to be the favoured pre-treatment option, although composting and anaerobic digestion (Chapter 2) may also be used in some circumstances. There are good arguments in favour of the incineration of combustible waste, especially when coupled to combined heat and power systems and effective control of waste incineration gases. However, if incineration is chosen as a pre-treatment option, then a further question arises; namely, should the bottom ash (approximately 10% by weight of the original waste) and fly-ash be disposed to monofill, or should it be codisposed? The answer to this question may be important in determining whether the bioreactor or pre-treatment scenarios predominate; It is important to recognise that the answers are, to an extent, mutually exclusive as the incineration of MSW removes the "medium" for effective co-disposal (See 7.4).

While the incineration of MSW will remove most (but not all) of the organic material, there are still likely to be problems associated with high levels of chloride, metals, and sulphate that would make sustainable disposal difficult to achieve. The ash from incineration (including the fly ash) could be disposed to a containment landfill that was dedicated to ash disposal, and in which liquid infiltration was prevented. But as discussed above, such containment systems must be expected to fail **ultimately**; in the absence of any attenuating mechanisms during storage, environmental pollution will result. This is not a sustainable option, as safe reassimilation into the environment will not be possible. As an alternative, the ash could be stabilised through processes such as solidification prior to landfill. Yet even after solidification, there is a risk that some toxic species will remain mobile, although the risk will be considerably less than monodisposal of untreated ash. Further treatment processes such as solidification would also significantly increase the cost of disposal. An alternative which does not appear to have been considered so far is monodisposal in a containment cell with high-rate liquid addition, collection, treatment (including pH modification to enhance leaching) and recirculation, such that upon liner failure, toxic elements of concern have been reduced to acceptable levels. Co-disposal options are discussed below.

Whether pre-treatment is effected by incineration, biological processing (such as composting), or by other means, the subsequent disposal to landfill will require careful management and control such that the risks to the environment are acceptable. The effort required for effective control will depend upon the waste stream and the pre-treatment process, and will be a significant factor (but one of many factors) in determining the most appropriate waste management option.

7.4   Hazardous waste disposal

Because radioactive waste decays with time but (in the absence of attenuating factors) toxic waste does not necessarily lose its toxicity, the safe disposal (as differentiated from treatment) of hazardous waste is potentially more difficult to achieve. According to Gera (1988) "containment is effective with radioactive materials or toxic substances that lose their toxicity with time, but in the case of stable toxic materials the result of any disposal practice is always dispersion in the environment". Recognising this, the best practice for hazardous waste disposal will always be one that attenuates the hazardous nature and operates as a treatment stage rather than relying on long-term storage.

According to Chapman and Williams (1988), the hazards involved in the disposal of low-level radioactive waste (LLW) are unlikely to be any more significant than those associated with the disposal of toxic MSW (with or without incineration as a pre-treatment) to monolandfill, yet the requirements for disposal of the latter regarding site selection, long term management and security are considerably less than for LLW. A comparison of waste toxicity has been shown to be difficult, not least because of the lack of data pertaining to toxic waste streams, but according to Chapman and Williams (1988) "it is likely that a successful analysis of comparative hazards would show that many persistent toxic wastes such as heavy metals, arsenic, asbestos, many organic chemicals such as PCB's and other cholorinated hydrocarbons, etc, are considerably more hazardous than some categories of radioactive waste, over which it is considered essential to exercise very much stricter controls, and which in any case decay quite rapidly". The application to toxic waste of disposal costs applied to LLW would probably effectively preclude monodisposal as a disposal option for such waste.

Within Europe, co-disposal (Chapter 3) will be banned, except at those sites with current co-disposal licenses, if the proposed Landfill Directive is implemented in its current format, while in the USA, the disposal of hazardous waste to landfill is not allowed without prior treatment. The banning of co-disposal, under these circumstances seems to be ill-advised, and born out of a short term attitude. Although the science and control of co-disposal is not **fully** understood, this should only serve to encourage further research rather than result in its banning. For co-disposal to be feasible, the waste within the landfill must be biologically active, and the more active the bacteria within the waste, the more efficient will be the attenuation and degradation of the added hazardous waste. However, even though the principle of co-disposal is sound, and data have been presented that show leachate from co-disposal sites to be similar to leachate from sites accepting domestic waste only, there is still much opposition to it's continued practice. Under conditions of optimum waste degradation and methane production, leachate and waste pH will remain at or near neutrality and is likely to do so until the remaining waste is effectively inert. Thus heavy metals are likely to remain immobile as insoluble salts (e.g. carbonates and sulphides). As most heavy metals were extracted from relatively inert material, their return to the land within a stabilised landfill could be considered as a suitable waste disposal option especially as it is an option over which some control can be exercised. If the bioreactor landfill approach were to prove successful, then the potential for the co-disposal of industrial waste including incinerator ash, would be significantly increased. This may be particularly relevant if MSW incineration increases and there is a need to dispose safely of large amounts of incinerator ash, although, as described above, the two are mutually exclusive: for if the waste is incinerated, then it will not be available to be used as a co-disposal "medium". Co-disposal of the ash may provide the means to attenuate the toxic species within the ash, and where this does not occur - such as with the chloride ion, then leachate control and treatment in a bioreactor landfill, utilising liquid recycle after further treatment, may provide the most effective means of treating the waste. It certainly represents an option where control can be effected, and does not rely solely on the integrity of long-term containment systems.

7.5   Conclusion

Both the European 5th Action Programme and The RCRA in the USA regard landfill as the last option within a waste management hierarchy. However when properly managed and properly controlled, landfill will continue to provide an effective and cost-efficient means of waste disposal for many years to come. Indeed, landfill is an **essential** part of an integrated waste management strategy, without which effective waste management would not be possible. Although many in the waste management industry believe that landfill can be effectively managed to prevent or minimise environmental pollution to an accceptable level, poor practices of the past and a lack of data in support of landfill as an effective waste treatment option do not encourage the continued development of the landfilling of waste in the way that it is currently undertaken in many parts of the world.

The potential for environmental pollution caused by the landfilling of waste is great, and polluting events caused by the migration of landfill gas and leachate have occurred too frequently in the past. But as identified above, landfill is an essential part of an integrated waste management system, and we must design, control, and operate landfills in such a way that the associated risks are recognised and effectively managed. The development of the sustainable landfill is the ultimate objective. This will be very difficult to achieve, but in striving, we will develop and operate landfills in a way that is more appropriate to 20th and 21st centuries.

Through research and development, our understanding of the processes involved in the production, properties, migration, attenuation and control of landfill gas and leachate have increased considerably in the past decade, such that the risk of pollution from a modern landfill has significantly decreased, provided that landfill activities are conducted in a well planned, operated and controlled manner. When waste is incinerated, the gaseous emissions are reassimilated into the environment. With appropriate emission control, the associated risks will be acceptable. The challenge for landfill disposal is that of facilitating eventual reassimilation of the emplaced waste into the environment, at acceptable risk.

However, although much research has already been conducted, there is a need for yet more, and there is an especial need to examine the optimisation of landfill reactions for enhanced waste stabilisation on a large-scale. There are still relatively few large-scale studies of the bioreactor-type landfill, and until meaningful studies have been conducted, it is difficult to anticipate to what extent waste stabilisation may be increased, and what problems will arise. The concept of the landfill as a treatment process (whether for treated or untreated waste), which will allow eventual reassimilation into the surrounding environment, seems to offer the best chance of truly sustainable waste management, for it does not rely on the long-term integrity of containment systems. However, the data that will be required to effectively assess this concept will take many years to accumulate. If alternative disposal methods are adopted in the meantime, it may be very difficult to change at some time in the future.

The cost of undertaking research at the scale of an operating landfill is great, and it would seem appropriate for this research to be undertaken as part of an international collaborative programme, where the costs and the results are shared between the nations.

One of the problems with such collaboration, is that within national scientific communities there are many who have a rather bigoted view on what constitiutes a sustainable landfill. i.e. they define the sustainable landfill in terms of design and operational procedures, and irrespective of local circumstances regard a particular approach (e.g. bioreactor or pre-treatment) as the only way forward. There will be circumstances under which the bioreactor approach will be most appropriate, and those under which pre-treatment is more sustainable; a risk assessment for a proposed development would determine the method of choice. The aim of international collaborative research should be to determine optimum conditions for each method, and to be able to provide data that would allow effective risk assessment to be undertaken.

A further obstacle to the development of sustainable landfilling practices is the reliance on a strict waste management hierarchy (Chapter 1), that places landfill as a last resort measure. Waste mimimisation and recycling should, under most circumstances, remain the preferred waste management options, but once these possibilities have been exhausted, the choice of treatment/disposal route should be on the basis of the Best Practicable Environmental Option. This will equate with the most sustainable option. Landfill will almost certainly constitute BPEO for some waste streams, and for these wastes, landfill should be used as the preferred management option.

There is no such thing as a "no risk" waste disposal operation, and a sustainable landfill cannot be defined in terms of its design and operation. There is no answer to whether pre-treatment or bioreactor approaches to landfill are the most sustainable. We should work towards gaining the knowledge and understanding that, at the planning stage, will allow informed decisions to be made on the least risk, most sustainable option. The concept that containment of any description will ultimately fail has been emphasised throughout this text. It is up to those in the landfill industry, and others with vested interests (e.g. national governments) to prove the effectiveness of landfill as a waste treatment option, which when after appropriate waste minimisation, re-use and recycle, can be used for the safe, controlled and cost-effective disposal of waste, in which the potential for environmental pollution is reduced to an acceptable level, and which ultimately accepts the reassimilation of the wastes into the surrounding environment.

REFERENCES

Andersen, T.V., Holm, P.E. and Christensen, T.H. (1991) Heavy metals in a landfill leachate pollution plume. Proc. 8th International conference, *Heavy metals in the environment*, 16 -20 September.

Anon. (1982) Gas generation from landfill sites. *County Surveyors Society*, UK. Committee No.4, Activity group No. 7. 40pp.

Anon. (1992a) Review of the environmental impact of recycling. *Warren Spring Laboratory report* LR511, UK.

Anon. (1992b) Counting the costs of leaking landfills. *ENDS Report* **205**, p 11.

Anon. (1992c) Landraising - Recent cases in Dorset and Humberside. *Waste Planning*, September issue, 26-28.

Anon. (1993a) Warmer Information sheet - Refuse-derived fuel. *Warmer Bulletin*, Nov. issue.

Anon. (1993b) Down to earth composting - of municipal green waste. Produced by the IWM scientific and technical committee. Institute of Wastes Management, Northampton, UK.

Anon. (1993c) Water pollution incidents the highest ever. *ENDS Report* **225**, 6-7.

Anon. (1994a) *Waste: a game of snakes and ladders* - A benchmarking report on waste and business strategy. Biffa Waste Services. UK.

Anon. (1994b) EC ban on co-disposal "would double" industry costs. *ENDS Report* **228**, p38.

Anon. (1994c) Landfill clean-up to cost "millions". *ENDS Report*, **229**, 12-13

Anon. (1994c) Revised strategy announced for renewable energy. *REview special issue*, August 1994, DTi, U.K.

Ashbee, E and Fletcher, I. (1993) Reviewing the options for leachate treatment. *Wastes Management*, Aug, 32-33

Augenstein, D. (1990) Greenhouse effect contributions to United States landfill methane. *Proc. GRCDA 13th Annual Landfill Gas Symposium*, March 27-29. GLFG No 0618, GRCDA, Silver Spring, MD. USA.

Baldwin, G. and Scott, P.M. (1991) Investigations into the performance of landfill gas flaring systems in the UK. In *Sardinia 91, Proc. Third International Symposium*, Vol 1. Grafiche Galeati, Imola. 301-312

Bardos, R.P. (1992) Composting studies using separated fraction from urban waste, from the Byker plant, Newcastle, United Kingdom. *Acta Horticulturae,* 302, 125-134.

Barlaz, M.A., Milke, M.W. and Ham, R.K. (1987) Gas production parameters in sanitary landfill simulators. *Waste Management Research*, **5**, 27

Barlaz, M.A., Ham, R.K. and Schaefer, D.M. (1989) Inhibition of methane formation from municipal refuse in laboratory-scale lysimeters. *Appl. Biochem. Biotechnol.* **20**, 181

Barlaz, M.A., Ham, R.K. and Schaefer, D.M. (1990) Methane production from municipal refuse: A review of enhancement techniques and microbial dynamics. *CRC Critical Reviews in Environmental control*, **19**(6),557-584.

Belevi, H. and Baccini, P. (1989) Long term assessment of leachates from municipal solid waste landfills. In *Sardinia 89. Proc. Second international landfill symposium,* Vol 1. CIPA, Milan. XXXIV,1-8.

Bishop, W.D., Carter, R.C. and Ludwig, H.F. (1966) Water pollution hazards from refuse - produced carbon dioxide. *J. Wat. Poll. Control. Fed.* **38**(3), 328-329.

Blake, D.R. and Rowland, F.S. (1988) Worldwide increase in tropospheric methane, 1978-1988. *J. Atmos. Chem.* **4**, 43-62.

Bratley, K. and Khan, A.Q. (1989) Waste management disposal methods: Landfill 2000. Presented at A.M.A. Environment conference *Caring for the future.* Newcastle-upon-Tyne. 10-12 July.

Buivid, M.G. (1980) Laboratory simulation of fuel gas enhancement from municipal solid waste landfills. Dynatec R&D Co., Cambridge, MA.

Buivid, M.G. et al (1981) Fuel gas enhancement by controlled landfilling of municipal solid waste. *Resource, Recovery and Conservation Management,* **6**, 3.

Buswell, A.M. and Hatfield, W.D. (1939) Anaerobic fermentations. *Illinois State Water Survey Bulletin,* **32**, 1-193.

Campbell, D.J.V (1989) Landfill gas migration, effects and control. In *Sanitary landfilling - Process, technology and environmental impact.* (Christensen,T.H., Cossu,R. and Stegmann, R. eds.) Academic Press. ISBN 0 12 174255 5. pp 399-424.

Campbell, D.J.V. (1991a) Landfill gas - under control? in *Proc. Harwell Waste Management Symposium* - Challenges in waste management. Harwell Laboratory, Oxon, UK.

Campbell, D.J.V. (1991b) The monitoring and control of landfill gas migration. in *Proc. Symp. Containment of pollution and redevelopment of closed landfill sites,* Leamington Spa, Construction Marketing, 2.2/1 - 2.2/11.

Campbell, D.J.V. (1992) Landfill - a major role to play in waste -and cost - containment. *NAWDC News,* Oct, pp 14,15 and 18.

Campbell, D.J.V. and Croft, B.C. (1991) Landfill gas enhancement - Brogborough test cell programme. In *Proc. Landfill gas: Energy and Environment '90* (Richards,G.E. and Alston,Y.R. eds.). Harwell Laboratory, Oxon, U.K.

Canter, L.W., Knox, R.C. and Fairchild, D.M. (1987) *Groundwater quality protection.* Lewis Publishers Inc. USA.

Card,G.B. (1992) Development on and adjacent to landfill. *J. Inst. Wat. Env. Man.* **6**(3), 362-371.

Carpenter, L.V. and Setter, L.R. (1940) Some notes on sanitary landfills. *Am. J. Public. Health,* **30**, 385-393

CEC (Council of European Communities) (1980) Directive on the protection of groundwater against pollution caused by certain dangerous substances. (80/68/EEC), Official Journal, L20, 43-47.

CEC (Commission of the European Communities) (1991) Proposal for a Council Directive on the Landfill of Waste (91/C190/01) *Official Journal of the European*

Communities, 22 July 1991, C190/1-18. Amended proposals COM(93)275. CEC, Brussels.

CEC (Commission of the European Communities) (1992a) Towards Sustainability-A European Community Programme of Policy and Action in Relation to the Environment and Sustainable Development, COM(92)23 final - vol. II. CEC, Brussels

Chandler, J.A., Jewell, W.J., Gossett, J.M., van Soest, P.J. and Robertson, J.B. (1980) Predicting methane fermentation biodegradability. *Biotechnology and Bioengineering Symposium*, **10**, 93-107.

Chapman, N.N. and Williams, G.M. (1988) Hazardous and radioactive waste management: a case of dual standards. In *Land disposal of hazardous wastes* (Gronow, J.R., Schofield, A.N. and Jain, R.K.), Ellis Horwood Ltd, Chichester, U.K. pp 259-268.

Cherry, J.A., Gillham, R.W. and Barker, J.F. (1984) Contaminants in groundwater: Chemical processes. In *Groundwater contamination* (studies in Geophysics, National Academy Press, Washington, USA.

Chian, E.S.K. and De Walle, F.B. (1977) *Evaluation of leachate treatment* Vol 1. Cincinatti EPA - 600/2-77-186a.

Chian, E.S.K. and De Walle, F.B. and Hammerberg, E. (1977) Effect of moisture regime and other factors on municipal waste stabilisation. In *Management of gas and leachate in landfills. Proceedings of the third annual Municipal solid waste research symposium.* EPA-600/9-77-026, USEPA, Cincinatti, OH. pp73-86

Christensen, T.H., Kjeldsen, P., Lyngkilde, J. and Tjell, J.C. (1989) Behavior of leachate pollutants in groundwater. In *Sanitary landfilling - Process, technology and environmental impact.* (Christensen,T.H.,Cossu,R. and Stegmann,R. eds.) Academic Press. ISBN 0 12 174255 5. pp 465-481.

Chynoweth, D.P. and Legrand, R. (1988) Anaerobic digestion as an integral part of municipal waste management. In *Landfill gas and anaerobic digestion of solid waste* (Alston, Y.R. and Richards, G.E.), pp 467-480, UKAEA, Harwell Laboratory, Oxfordshire, UK.

Clay, A. and Norman, T. (1989) Landfill wastes - it's a gas. *Gas. Eng. Manage.* **29** (11-12), 314-320, 322-325.

Coggins, P.C. (1993) The important option of waste minimisation. *Wastes Management Proceedings*, April, 9-12.

Coopers and Lybrand (1993) *Landfill Costs and Prices: Correcting Possible Market Distortions.* HMSO, London.

COSHH (1988) United Kingdom Health and Safety Executive. The Control of Substances Hazardous to Health 1988. Statutory Instrument 1988, No 1657, HMSO London.

Dass, P.,Tamke, G.R. and Stoffel, C.M. (1977) Leachate production at sanitary landfill sites. *J. Env. Eng. Div., Proc. ASCE,* **103**(EEG), 981-988.

De Baere, L. and Six, W. (1988) Dry anaerobic digestion of agro-industrial wastes. In *Landfill gas and anaerobic digestion of solid waste* (Alston, Y.R. and Richards, G.E.), pp 545-557, UKAEA, Harwell Laboratory, Oxfordshire, UK.

DoE (Department of the Environment) (1978) *Cooperative programme of research on the behaviour of hazardous wastes in landfill sites.* Final report of the Policy review committee., (Chairman J.Sumner) HMSO, London.

DoE (Department of the Environment) (1986) *Landfilling Wastes, Waste Management Paper No 26, A Technical Memorandum for the Disposal of Wastes on Landfill Sites.* HMSO, London

DoE (Department of the Environment) (1992a) *Landfill Gas. Waste Management Paper No 27, A Technical memorandum providing guidance on the monitoring and control of Landfill Gas.* HMSO, London

DoE (Department of the Environment) (1992b) *The technical aspects of controlled waste management: Understanding landfill gas.* Report CWM/040/92.

DoE (Department of the Environment) (1993a) *Externalities from Landfill and Incineration* - A Study by CSERGE Warren Spring Laboratory and EFTEL. HMSO, London

DoE (Department of the Environment) (1993b) *The potential for woodland establishment on landfill sites.* (ref CWM 054/92), HMSO, London.

DoE (Department of the Environment) (1993c) *The technical aspects of controlled waste management - Understanding landfill gas.* Report No. CWM/040/92

DoE (Department of the Environment) (1993d) *Waste Management Paper 26A: Landfill Completion, A technical memorandum providing guidance on assessing the completion of licensed landfills.* HMSO, London.

DoE (Department of the Environment) (1994a) *Waste Management Paper No. 26F: Landfill Co-Disposal.* A Draft for Consultation. DoE

DoE (Department of the Environment) (1994b) *Sustainable Development: The UK Strategy.* Cm 2426. HMSO, London.

DoE (Department of the Environment)(1995a) *A Waste Strategy for England and Wales - Consultation Draft.* HMSO, London.

DoE (Department of the Environment) (1995b) *Waste management paper 26b: Landfill design, construction and operational practice.* A draft for consultation. UK DoE.

Doedens, H. and Cord-Landwehr, K (1989) Leachate recirculation. In *Sanitary Landfilling: Process, Technology and Environmental Impact* (Christensen, T.H., Cossu, R. and Stegmann, R. eds.) Academic Press, London.

Eden, R. (1994) Landfill gas collection techniques and technology. In *Landfill gas Seminar : Energy from landfill gas - making it work.* Seminar Prospectus, ,17 March 1994, Solihull, U.K.

Edworthy, K.J. (1989) Waste disposal and groundwater management. *J. IWEM.* **3**, 109-115.

Ehrig, H.J. (1991) Control and treatment of landfill leachate - A review. In *1991 Harwell waste management symposium: Challenges in waste management.* Harwell Laboratory, Oxfordshire, UK.

Eliassen, R. (1942) Decomposition of landfills. *Am.J.Public Health*, **32**, 1029-1037.

Emberton, J.R. (1986) The biological and chemical characterisation of landfill. Proc. UK DEn/US OE Conf. Solihull, UK. OCT 28-31 p 150-163.

EMCON Associates (1975) Sonoma County Solid Waste Stabilisation study, EPA 530/SW 65d.1, PB 239 778. USEPA, Cincinatti, OH.

EMCON Associates (1981) State of the art in methane gas enhancement in landfills. Argonne National Laboratory Report no.ANL/CNSV-23, US Department of Energy.

Esmaili, H (1975) Control of Gas Flow from sanitary landfills. *J. Environ. Eng-Div, Am. Soc. Civ. Eng.*, **101** (EE4) 555-566.

ETSU (1990) *Fundamental Studies On Cellulose Degradation in Landfills*. Contractor Report ETSU B 1228. Harwell Laboratory, Oxon, UK.

ETSU (1991) Landfill Gas and the Environment. *Landfill Gas Trends* **3** Oct. Harwell Laboratory, Oxon, UK.

ETSU (1992) *Review of the technologies for monitoring the microbiology of landfill*. Contractor report ETSU B 1316. Harwell Laboratory, Oxon, UK.

ETSU (1993a) *The sustainable landfill: A feasibility study to assess the potential of developing a bioreactor cell rotational landfill*. ETSU report B/B3/00242/rep, Harwell Laboratory, Oxon, UK.

ETSU (1993b) *Landfill gas enhancement studies: The Brogborough test cells*. ETSU report B/B5/00080/rep, Harwell Laboratory, Oxon, UK.

Fairbank (1994) Recycling - the inevitable reality. *IWM Proceedings*, July, 22-24.

Farquhar, G.J. and Rovers, F.A. (1973) Gas production during refuse decomposition. *Water, Air, and Soil Pollution*, **2**, 483-495.

Fenn, D.G., Hanley, K.J. and Degeare, T.V. (1975) Use of the water balance method for predicting leachate generation from solid waste disposal sites. EPA-530/SW-168. US EPA, Cincinnati, Ohio,USA, 40pp.

Figueroa, R.A. and Stegmann, R. (1991) Gas migration through natural liners. In *Sardinia 91, Proc. Third International Symposium. Vol 1*. Grafiche Galeati, Imola. 167-177.

Fielding, E.R., Archer, D.B., Conway de Macario, E. and Macario, A.J.L. (1988) Isolation and characterisation of methanogenic bacteria from landfills. *Appl. Environ. Microbiol*, **54**, 835-836.

Foster, S.S.D.(1987) Fundamental concepts in aquifer vulnerability, Pollution risk and protection strategy. In *Vulnerability of soil and groundwater to pollutants*. (van Duijvenbooden,W. and van Waegeningh,H.G. eds.) pp 69-86.

Freeman, H.M. (1995) Pollution Prevention - A New Agenda. In *Industrial Pollution Prevention Handbook* (Freeman, H.M. ed.) pp 1-8. McGraw-Hill, USA.

Freeman, H.M. and Lounsbury, J. (1990) Waste Minimisation as a Waste Management Strategy in the United States. In *Hazardous Waste Minimisation* (Freeman, H.M. ed.) pp 3-14. McGraw-Hill, USA.

Freeze, R.A. and Cherry, J.A. (1979) *Groundwater*. Prentice Hall.

Gendebien, A., Pauwels, M., Constant, M., Ledrut-Damanet, M.-J., Nyns, E.-J., Willumsen, H.-C., Butson, J., Fabry, R. and Ferrero, G.-L. (1992) *Landfill gas from environment to energy.* CEC, Luxembourg. ISBN 92-826-3672-0

Gera, F. (1988) Modelling long term impacts of land disposal of hazardous waste. In *Land disposal of hazardous waste.* (Gronow,J.R., Schofield, A.N. and Jain,R.K. eds.) John Wiley and Sons, pp 93-102.

Ham, R.K. (1975) The role of shredded refuse in landfilling. *Waste Age*, **6**, 22.

Harmsen, J. (1983) Identification of organic compounds in leachate in a waste tip. *Water Res.* **17**, 699.

Harries, C.R. (1988) Landfill microbiology: work supported at Biotal by the department of Environment In *Landfill Microbiology: R & D workshop* (Lawson, P and Alston, Y.R. eds.) Harwell Laboratory, Oxon, UK. pp 150-176.

arris, R.C. (1988) Leachate migration and attenuation in the unsaturated zone of the triassic sandstones. In *Land disposal of hazardous waste.* (Gronow,J.R., Schofield, A.N. and Jain,R.K. eds.) John Wiley and Sons, pp 175-186.

Harris, R.C., Knox, K. and Walker, N. (1994) A strategy for the development of sustainable landfill design. *IWM Proceedings*, Jan, 26-29.

Haukohl, J. (1993) The experience of Denmark in waste to energy projects. *Wastes Management,* Dec. p9

Hoather, H.A. and Wright, P.W. (1989) Landfill Gas: Site licensing and risk assessment. In *Proc. Department of Energy conf. on landfill gas and anaerobic digestion of solid waste.* (Alston,Y.R. ed.) Chester, Oct 1988, Harwell laboratory, 100-128

Holmes, J (1993) Waste management practices in developing countries. *Wastes Management,* June, 8-14.

Institute of Wastes Management (1991) *Monitoring of landfill gas.* Available 9 Saxon Court, St Peter's Gardens, Northampton, NN1 1SX, UK.

Intergovernmental Panel of Climate Change (IPCC) (1992) *Climate change 1992. The supplementary report to the Intergovernmental Panel on Climate Change Scientific Assessment,* (Houghton, J.T., Callander, B.A. and Varney, S.K. eds.) Cambridge, UK. Cambridge University Press.

ISWA (International Solid Wastes Association Working Group on Waste Incineration) (1991) *Energy from waste state-of-the-art report,* ISWA, Malmo, Sweden.

ISWA (International Solid Wastes Association) (1992) *1000 Terms in Solid Waste Management* (Skitt, J. ed.). ISWA, Copenhagen.

Jenne, E.A. (1968) Control of Mn, Fe, Ni, Cu, and Zn concentration in soils and waters, significant roles of hydrous Mn and Fe oxides. In *Trace inorganics in water. ACS. adv. Chem. Ser.73.* Washington D.C. pp 337-387.

Jones-Lee, A. and Lee, G.F. (1993) Groundwater pollution by MSW landfill: Leachate composition, detection and water quality significance. In *Proceedings Sardinia 93, Fourth International Landfill Symposium*, S. Margherita di Pula, Cagliari, Italy, 11-15 October.

Kersey, J.D. (1994) Sustainable development and wastes management. *Wastes Management*, Jan, 48-52.

Klink, R.E. and Ham, R.K. (1982) Effects of moisture movement on methane production in solid waste landfill samples. *Resources and Conservation*, **8**, 29-41.

Knox, K. (1989) *A review of technical aspects of co-disposal.* A report prepared for the UK Department of the environment. No. CWM 007/89.

Knox, K. (1990a) A review of co-disposal. *Proc.1990 Harwell Waste Management Symposium.* Environmental Safety Centre, Harwell Laboratory.

Knox, K. (1990b) The relationship between leachate and gas In *Proceedings of international conference. Landfill gas: Energy and environment '90*, Bournemouth, Oct.

Knox, K. (1991) A review of water balance methods and their application to landfill in the UK. Report prepared for the Department of the Environment, No. CWM 031/91.

Knox, K. (1992) The role of research in controlling co-disposal in the UK. In *Proc. 1992 Harwell Waste Management Symposium - New Developments in Landfill.*

Knox, K. (1995) A review of the Brogborough test-cell project. In *Landfill gas microbiology workshop prospectus*, St. John's Swallow Hotel, Solihull, UK (organised by ETSU, Harwell Laboratory, Oxon, UK.)

Knox, K. and Gronow, J. (1993) A review of landfill cap performance and it's application for leachate management. In *Proceedings Sardinia 93, Fourth International Landfill Symposium*, S. Margherita di Pula, Cagliari, Italy, 11-15 October, pp 207-223.

Kornberg, J.F., Von Stein, E.L. and Savage, G.M. (1993) Landfill mining in the united states: An analysis of current projects. In Proceedings Sardinia 93, Fourth International Landfill Symposium, S. Margherita di Pula, Cagliari, Italy, 11-15 **October.**

Laurijssens (1993) Waste policy in the Netherlands. *Wastes Management.* Nov. issue, 12-14.

Lee, G.F. and Jones-Lee, A. (1993) Landfill and groundwater pollution issues: "Dry tomb" vs wet-cell landfills. In Proceedings Sardinia 93, Fourth International Landfill Symposium, S. Margherita di Pula, Cagliari, Italy, 11-15 **October.**

Leckie, J.O., Pacey, J.G. and Halvadakis, C. (1979) Landfill management with moisture control. *J. Env. Eng. Division, ASCE*, **105**,(EE2), 337-335.

Loxham, M. (1993) The design of landfill sites - some issues from a European perspective. In *Landfill tomorrow - Bioreactors or storage.* Proceedings of seminar held at Imperial College of Science, Technology and medicine, London. pp7-12.

Lu, J.C.S., Eichenberger, B. and Stearns, R.J. (1985) Leachate from municipal landfills-Production and management. Pollution technology review No.119, Noyes Publication, Park Ridge, New Jersey, USA.

LWRA (London Waste Regulation Authority) (1995) Today's Waste, Tomorrow's Resources - The Waste Management Plan for Greater London 1995 - 2015 (Draft Document). LWRA, London.

Lyngkilde, J. and Christensen, T.H. (1992) Fate of organic contaminants in the redox zones of a landfill leachate pollution plume. *J. Contaminant Hydrology*, **10**, 291-307.

Massman, J.W., Moore, J.W. and Sykes, R.M. (1981) Development of computer simulation for landfill methane recovery, *Technical Report ANL/CNSV-26. Argonne Nat. Lab.*, US Department of Energy.

Maurer, R.W. (1993) Landfill tomorrow: a paradigm shift from storage to bioreactors. In *Landfill tomorrow - Bioreactors or storage*. Proceedings of seminar held at Imperial College of Science, Technology and medicine, London. pp1-6

Meadows, M., Brown, K. and Lawson, P. (1994) Making landfill gas work: A summary of the Department of Trade and Industry Energy from landfill gas programme 1993-1998. In *Landfill gas Seminar : Energy from landfill gas - making it work*. Seminar Prospectus, ,17 March 1994, Solihull, U.K.

McEntee, J.M. and Jelley, J.C. (1991) Legal implications of methane generation from old landfill sites. In *Methane-Facing the Problems*. Second Symp. and exhibition. Nottingham 26-28 March. pp 6.4.1-6.4.11.

National Rivers Authority (1992) *Policy and practice for the protection of groundwater*. ISBN 1 873160 37 2.

North West Waste Disposal Officers (Liners sub-group) (1986) *Guidelines on the use of landfill liners*. Lancashire Waste Disposal Authority.

North West Waste Disposal Officers (1991) *Leachate management Report*. Lancashire Waste Disposal Authority.

Op den Camp, H.J.M., Verhagen, F.J.M., Kivaisi, A.K., de Windt, F.E., Lubberding, H.J., Gijzen, H.J. and Vogels, G.D. (1988). Effects of lignin on the anaerobic degradation of (ligno)cellulosic wastes by rumen microorganisms. *Applied microbiology and Biotechnology*, **26**, 56-60.

Pacey, J.G. (1986), 'The factors influencing landfill gas production *Proc. UK D.En/US DOE Conf. Solihull*, UK OCT 28-31. pp 51-59.

Pacey, J. and Karpinski, G. (1980) Selecting a landfill liner, *Waste Age*, **11**(7), 26-28.

Peters, T. (1993) Chemical and physical changes in the subsoil of three waste landfills. *Waste Man. Res.*, **11**, 17-25.

Petts, J.I. (1993) Risk assessment, risk management and containment landfill. In 1993 Harwell waste management symposium: Options for landfill containment. AEA Technology, Harwell, Oxofordshire, UK.

Petts, J.I. and Eduljee, G. (1994) *Environmental impact assessment for waste treatment and disposal facilities.* John Wiley and Sons Ltd, England.

Pfeffer, J.T. (1974) Temperature effects on anaerobic fermentation of domestic refuse. *Biotech. Bioeng.*, **16**,771-787.

Philpott, M.J., Reid, R.C., Davies, J.N., Last, S.D. and Boldon, J.M. (1992) Environmental control measures at Greengairs landfill site. *J.Inst.Wat.Env.Man.* **6**(1), 38-47.

Pohland, F.G. (1975) *Sanitary landfill stabilisation with leachate recycle and residual treatment.* EPA grant R-801397, Georgia Institute of Technology, Atlanta, USA.

Pohland, F.G. (1989) Leachate recirculation for accelerated landfill stabilisation. In *Sardinia '89: Proceedings of the second international landfill symposium,* Porto Conte. XXX, pp1-8, CIPA, Milan

Pohland, F.G. (1991) Fundamental principles and management strategies for landfill co-disposal practices. In *Sardinia 91, Proc. Third International Symposium. Vol 2.* Grafiche Galeati, Imola. pp1445-1460.

Quarrie, J (1992) *Earth Summit '92* - The United Nations Conference on Environment and Drevelopment, Rio de Janeiro. Regency Press, London

RCEP (Royal Commission on Environmental Pollution) (1976) *Air pollution control: an integrated approach.* Fifth Report, Cmnd 6371. HMSO, London.

RCEP (Royal Commision on Environmental Pollution) (1984) *Tackling pollution - Experience and prospects.* 10th report, Cmnd 9149

RCEP (Royal Commission on Environmental Pollution) (1988) *Best Practicable Environmental Option.* 12th Report, Cm 310. HMSO, London.

RCEP (Royal Commission on Environmental Pollution) (1993) *Incineration of Waste.* 17th Report, Cm 2181. HMSO, London

Rees, J.F. (1980) The fate of carbon compounds in the landfill disposal of organic matter. *J.Chem.Tech.Biotechnol.* **30**,161-175.

Rees, J.F. (1982) Landfill management and leachate quality. *Effluent Water Treat.J.*, **22**,457.

Rees, J.F. and Grainger, J.M. (1982) Rubbish dump or fermenter? Prospects for the control of refuse fermentation to methane in landfills. *Process Biochemistry*, **17**(6), 41-44

Rettenberger, G. (1985) Gasformige emissiononen aus abfalldeponien im himblick auf umweltrelevante schadstoffe. (Gaseous emissions from waste landfills in regard to environmental relevant toxic compounds) In *GRCDA Landfill gas. Proc. 8th Int. Symp.* San Antonio, Texas. GRCDA-SWANA, PO Box 7129, Siverspring, Maryland, 20910, USA.

Richards, K.M. and Aithcheson, E.M. (1991) Landfilling in the greenhouse world. Landfill gas exploitation in the UK. in *Sardinia 91. Proc. Third International Landfill Symposium Vol 1.* Grafiche Galeati, Imola, Italy. pp33-44.

Robinson, H.D.(1990a) On site treatment of leachates from landfilled wastes. *J.IWEM.*, **4**(1), 78-89.

Robinson, H.D.(1990b) Leachate composition and treatment. *Proc. 1990 Harwell waste management symposium - The 1980's. A decade of progress?* Harwell Laboratory, Oxfordshire, UK.

Robinson, H. and Gronow, J. (1992) Groundwater protection in the UK: Assesment of the landfill leachate source term. *J.Inst.Wat.Env.Man.* **6**(2),229-236.

Robinson, H.D. and Maris, P.J. (1979) *Leachate from domestic waste: Generation, composition and treatment. A review.* Water Research Centre Technical Report TR 108, 38pp.

Robinson, H.D., Barr, M.J. and Last, S.D. (1992) Leachate collection treatment and disposal. *J.Inst.Wat.Env.Man.* **6**(3), 321-332.

Robinson, H.D., Barr, M.J., Formby, R. and Morag, A. (1993) Using reed bed systems to treat landfill leachate. *Wastes Management Proceedings*, April, 20-26.

Rovers, F.A., Tremblay, J.J. and Mooij, H. (1978) Procedures for landfill gas monitoring and control. Waste management report EPS 4-EC-77-4on the *Proc. International Seminar Fisheries and Environment.* Minister of Supply and Services, Montreal, Canada

Ruckelshaus, W.D. (1989) Toward a sustainable World. *Scientific American*, September pp166-175.

Rylander, H. (1993) MSW management, policies and strategies in Sweden. *Wastes Management*, Dec. pp13-15.

Savage, G.M. and Diaz, L.F. (1994) Landfill mining and reclamation. *ISWA Times.* **4**, 2-4

Scott, P.E. (1990) Microcomponents of landfill gas: Toxicity and corrosion problems. *Proc. Brunel Landfill Conference - Problems and Solutions.* Brunel University, Uxbridge, UK.

Scott P.E., Dent,C.G. and Baldwin,G. (1988) *The composition and environmental impact of household waste-derived landfill gas: Second report.* AERE-G-4436, UKAEA, Environment and Energy, Oxfordshire, England.

Segal, J.P. (1987) Testing large landfill sites before construction of gas recovering facilities. *Waste Man. Res.* **1** p201.

Senior, E. and Kasali, G.B. (1990) Landfill Gas in *Microbiology of Landfill Sites* (Senior,E. ed.) CRC press pp 113-158.

Senior, E. and Shibani, S.B. (1990) Landfill leachate in *Microbiology of Landfill Sites* (Senior,E. ed.) CRC Press. pp 81-112.

Seymour, K.J. (1992) Landfill lining for leachate containment. *J.Inst.Wat.Env.Man.* **6**(4), 389-396.

Stegmann, R. (1991) Role of sanitary landfilling in solid waste management. in *Sardinia 91, Proc. Third International Symposium. Vol 1.* Grafiche Galeati, Imola. pp3-13.

Stegmann, R. (1993) **Management of a dry landfill system** (DLS) In *Proceedings Sardinia 93, Fourth International Landfill Symposium*, S. Margherita di Pula, Cagliari, Italy, 11-15 October.

Thurber, J. and Sherman, P. (1994) Pollution prevention requirements in United States environmental laws. In *Industrial Pollution Prevention Handbook* (Freeman, H.M. ed.) pp27-49. McGraw-Hill, USA.

Tyson, R. (1992) prospectors mining profit from old landfills. *USA Today/International edition*, April 25, p5A.

USEPA (US Environmental Protection Agency) (1990a) *Meeting the Environmental Challenge:* EPA's Review of Progress and New Directions in Environmental Protection, EPA, 21K-2001.

USEPA (US Environmental Protection Agency) (1990b) *Air Emissions from Municipal Solid Waste Landfills - Background Information for Proposed Standards and Guidelines.* Draft Environmental Impact Statement, Office of Air Quality, Research Triangle Park.

USEPA (US Environmental Protection Agency) (1991) USEPA Solid waste disposal facility criteria; Final rule. 40 CFR parts 257 and 258, *Federal register,* 56 (196): 50978 - 51119.

USEPA (US Environmental Protection Agency) (1992a) *Characterisation of Municipal Solid Waste in the United States:1992 Update.* USEPA, Office of solid waste and emergency response. July. EPA/530-R-019.

USEPA (US Environmental Protection Agency) (1992b)*A groundwater information tracking system with statistical analysis capability GRITS/STST v4.2* U.S Environmental Protection Agency, Office of Research and Development, Office of Solid Waste, and region VII. EPA/625/11-91/002.

USEPA (US Environmental Protection Agency) (1994a) *Design, operation, and closure of municipal solid waste landfills.* EPA/625/R-94/008

van Santen (1993) Energy from waste: a perspective. *IWM Proceedings*, Jan, 11-19.

Walker, N. (1993) Landfill leachate control: A diagnosis and prognosis. *Wastes management proceedings*, Jan, 3-10.

Wallis, M. (1994) Waste incineration reassessed. *Warmer Bulletin*, 41, 18-19.

Wallis, M. and Watson, A. (1995) The push for waste incinerators in the EU and UK is challenged. *IWM Proceedings*. April issue, 6-7.

Wheeler, P.A., Border, D.J. and Riding, A.E. (1994) The market for composts and digestates from organic municipal wastes. Warren Spring Laboratory report No LR 1011. ISBN 085624 871 1.

Williams, G.M. and Hitchman, S.P. (1989) The generation and migration of gases in the subsurface. In *Methane-Facing the Problems*. Nottingham University 26-28 Sept pp 2.1.1 - 2.1.12.

Williams, G.M. and Aitkenhead, N. (1991) Lessons from Loscoe: the uncontrolled migration of landfill gas. *Quarterly J.Eng.Geol.* **24**,191-207.

Williams, G.M. Smith, B. and Ross, C.A.M. (1991) The migration and degradation of waste organic compounds in groundwater. *Adv.Org.Geochem.***19**, 531-543.

Zaoui, R. (1988) Valorga digestion process. In *Landfill gas and anaerobic digestion of solid waste* (Alston, Y.R. and Richards, G.E.), pp545-557, UKAEA, Harwell Laboratory, Oxfordshire, UK.

A PHONETIC STUDY OF
WEST AFRICAN LANGUAGES

A PHONETIC STUDY OF
WEST AFRICAN LANGUAGES

AN AUDITORY-INSTRUMENTAL SURVEY

BY

PETER LADEFOGED

Professor of Phonetics,
University of California, Los Angeles

SECOND EDITION

CAMBRIDGE

AT THE UNIVERSITY PRESS

1968

PUBLISHED BY THE SYNDICS OF THE CAMBRIDGE UNIVERSITY PRESS
Bentley House, P.O. Box 92, 200 Euston Road, London, N.W.1
American Branch: 32 East 57th Street, New York, N.Y. 10022

This edition © Cambridge University Press, 1968

First Edition published by the Syndics of the Cambridge University
Press in association with the West African Languages Survey, as
No. 1 of the West African Language Monograph Series   1964

Second Edition   1968

Standard Book No. 521 06963 7

Library of Congress Catalogue Card Number: 68-13541

First printed in Great Britain by Percy Lund, Humphries & Co. Ltd.,
London and Bradford

Reprinted in Great Britain by Lowe & Brydone (Printers) Ltd., London

CONTENTS

LIST OF PLATES

vii

FOREWORD

The present work, based as it is on materials from sixty-one West African languages, represents the first contribution of broad scope in the area of phonetic studies in Africa since the publication in 1933 of Westermann and Ward's *Practical Phonetics for Students of African Languages*. This latter work has been a landmark in its field and a fundamental element in the training of African linguistic fieldworkers since its first appearance. The remarkable advances both in our knowledge of substantive data and in phonetic research methods in the intervening period is reflected in the present work. It might appear at first glance as though the difference in method involves chiefly the replacement of auditory by instrumental methods. However, Dr. Ladefoged rightly stresses the continuing and fundamental role of the trained human ear alongside the instrumental methods in the very subtitle of his work, 'An Auditory-Instrumental Survey'. After all, we are concerned with the physical nature of sound in so far as differences are perceivable by the human ear and, as such, utilized in the phonetic systems of human languages.

It is clear that Dr. Ladefoged has made fundamental contributions to the subject of the phonetics of West African languages (e.g. his discovery that the sounds called labiovelars are produced in three different ways) and opens up new topics for further investigation. I believe, however, that Dr. Ladefoged's work is of more than purely African significance. His more general contributions lie, in my opinion, chiefly in two directions. By investigating phenomena not present in Western European languages, his work serves to broaden the general perspectives of phonetic research. The other main value may be seen in the variety of instrumental techniques which are here brought to bear. This allows an investigation of particular phenomena in which both the articulatory and acoustic sides receive their due. The fruitfulness of such a combined gennemic and acoustic approach may again be exemplified by reference to the investigation of labiovelars where the articulatory differences which distinguish the various types are deduced from pressure measurements in the relevant cavities, a result probably not attainable even by the most painstaking study of the acoustic phenomena in isolation.

Dr. Ladefoged's work is therefore a contribution to general phonetics, and hence to linguistic disciplines as a whole, as well as to African language studies. It illuminates the fact, often to be observed in the history of linguistics, that scientific studies of one language, or a group of languages, can be expected to offer insights and perspectives of a more general significance for students of language.

JOSEPH H. GREENBERG
STANFORD UNIVERSITY, CALIFORNIA *Chairman of the Council*
West African Languages Survey

ACKNOWLEDGEMENTS

A great deal of this monograph is not my own original work. Not only have I relied on the references cited in the bibliography, but I have also had a considerable amount of help from personal communications and discussions, particularly with Robert G. Armstrong, the Field Director óf the West African Languages Survey, John Spencer, who also had all the problems of seeing the work through the press, Peter Cockshott, who supervized the radiology, and, among many other linguists whom I hope I have acknowledged at appropriate places in the text: Gilbert Ansre, David Arnott, Elizabeth Dunstan, Carl Hoffmann, Maurice Houis, Cath McCallien, Gabriel Mannesy, Alfred Opubor, Walter Pichl, M. Sauvageot, Berthe Siertsema, John Stewart, William Welmers, and André Wilson. In addition, I am, of course, especially indebted to Joseph Greenberg, the Chairman of the West African Languages Survey Council, not only for his continued interest and advice, but also for kindly contributing a foreword.

P. L.

UNIVERSITY OF CALIFORNIA,
LOS ANGELES. *August 1963*

NOTE ON SECOND EDITION

Apart from minor corrections, this edition is the same as the first. I would like to express my thanks to a number of reviewers, particularly J. Carnochan, J. Kelly, W. Samarin, Berthe Siertsema and Elizabeth Uldall. I have corresponded with all these scholars, and I have amended all the errors they were able to point out to me. In a few cases I have not made the changes suggested by J. Carnochan and J. Kelly, because they are either factually incorrect, or their phonetic observations do not agree with my own.

P. L.

UNIVERSITY OF CALIFORNIA,
LOS ANGELES. *February 1967*

INTRODUCTION

This monograph is based on data from sixty-one languages. This is, of course, an inadequate sample of the languages of West Africa, which are both very numerous and extraordinarily diverse in character. Nobody knows exactly how many there are. Westermann and Bryan (1952) list about 173 different languages or dialect clusters; and to my personal knowedge many of their dialect clusters consist of several quite distinct languages. Answers to a questionnaire issued before a recent conference (Spencer, Ed., 1963) indicate that there may be nearer 300 different languages in Sierra Leone, Ivory Coast, Ghana and Nigeria. If we add in the languages spoken in Senegal, Gambia, Portuguese Guinea, Guinea, Mali, Liberia, Upper Volta, Niger, Togo, Dahomey and the part of the Cameroun Republic which is usually considered to be in West Africa, the figure for the total number of languages would be enormously increased, almost certainly to well over 500. In addition the area contains languages from at least two totally unrelated language stocks (Greenberg, 1963). Consequently at this stage of our ignorance, it would be foolish to try to predict which languages are in any way representative.

The languages discussed here include most of those used for official purposes such as broadcasting in Senegal, Sierra Leone, Ghana and Nigeria. In my capacity as a Fellow on the West African Languages Survey I worked at the four Universities represented on the Survey Council (University of Dakar, Senegal; University College, Fourah Bay, Sierra Leone; the University of Ghana; and the University of Ibadan, Nigeria). I was able to investigate nearly all the languages spoken by large groups of educated people in these countries. In addition, I considered a number of the less well known languages spoken by students in the universities in Ghana and Nigeria. The geographical range of all the languages investigated is indicated by the map (Fig. 1) which shows the hometowns of most of the main informants. The typological spread is shown by Appendix A which is arranged in accordance with the classification suggested by Greenberg (1963).

Apart from a few major languages such as Hausa and Fula, which are spoken over very wide areas, only one or two informants were used for each language. This is a serious limitation; it may mean that any particular piece of data is really applicable to no more than a single speaker, and may not be typical of the language as a whole. The informants were, however, carefully chosen: none of them had any obvious speech defect or unusual vocal organs; and they all claimed to be accepted as normal speakers by the people of their hometowns. So the general description of each sound is probably valid, although not too much emphasis should be placed on the particular values shown in the tables and illustrations. A list of the principal informants, their occupations and their home localities is given in Appendix A.

The material in this book is arranged in two parts. The first is a description of some of the phonetic elements that are used to cause differences in lexical or grammatical meaning in the languages under consideration. I have not attempted to describe every sound in each language. The student of African languages is fortunate in having available an excellent introduction to the phonetic basis of his subject. Westermann and Ward (1933) describe all the features of speech that are common to most of the languages of the world; and they also give an account of many other phenomena which are frequent in African languages. In this book I have concentrated on the more unusual sounds which

Westermann and Ward have either omitted or not described very fully. I need hardly add that I am not concerned with phonetic description for its own sake; there is no discussion of oddities of sound which do not have a linguistic function in any language. The intention is simply to provide linguists analysing these and similar languages with a fairly detailed account of a number of the phonetic events which are linguistically significant.

The second part of the book consists of the appendices, one of which lists some of the linguistic oppositions in each language; usually there is an example of the contrast between each of the consonant phonemes. The phonological analyses underlying these lists are not all of the same standard. The lists are derived from examples collected in the course of the phonetic investigations which form the first part of the book. Whenever possible the work with the informants consisted of eliciting contrasts reported in the literature or in personal communications from my colleagues. But no phonological information was available for a number of the languages; and in these cases I had to make a rudimentary analysis myself. All these analyses are naturally very tentative. The consonant contrasts are usually fairly reliable; but some of the vowel phonemes may be misrepresented; and the tone marking in most cases is simply a record of what was said on a particular occasion, usually marked in terms of the three possibilities: high, ' mid (unmarked), and low `.

In both parts of the book I have followed a strict rule of leaving out of consideration all contrasts which I have not heard myself from an informant, and which are known to me only through the literature or by means of tape recordings made by other people. This means that I have not been able to discuss or list in either part of the book a number of sounds which are mentioned by well known authorities; but it also follows that all the phonetic data conform to the same standard.

Most of the problems are stated in terms of differences between phonemes (and occasionally between allophones). This is only to provide an easily understood means of labelling the phonetic events. Some of the oppositions discussed are not contrasts in single sounds, but are contrasts in pieces which extend over longer spans, and could therefore be considered as properties of syllables or larger phonological units; and equally some of the linguistic oppositions could be more economically described in terms of the smaller distinctive features into which the phonemes may be divided. This is irrelevant to the object of the current work, which is simply to establish what happens when certain words and phrases are said. I am concerned with the nature of the phonetic events which can be correlated with lexical or grammatical contrasts. I would like to discuss the organization of these events in terms of phonological systems and structures; but for the moment I have avoided nearly all such discussions in order to keep this survey of the phonetic bases of so many languages within reasonable limits.

For similar reasons a traditional approach is used in the presentation of the phonetic material. The descriptive terms are used in accordance with the definitions in Abercrombie (forthcoming); the few neologisms which I have been forced into are discussed as they occur. The data might well have been organized differently, perhaps in terms of distinctive features, or acoustic parameters suitable for synthesizing speech by rule. Those who wish to consider the events in these and other ways (which are certainly more appropriate for many linguistic purposes) can easily do so on the basis of the data provided. But I hope that the present formulation will ensure that the work is useful to as many linguists as possible.

Whenever a word in an African language is cited, a gloss of some kind is also given. All such glosses are very rough, and are intended simply as guides which may help the reader identify forms in languages which he knows or can obtain from informants. In many cases I was unable to identify the particular item of the general kind specified by the informant. Such glosses appear with a prefixed asterisk; thus (*fruit) and (*fish) denote particular kinds of fruit and fish. Proper names are also cited with a prefixed asterisk. It must be remembered that in many of the languages I was working through French with informants who also had only a limited knowledge of that language; and often I may not have chosen the appropriate English gloss of the French word or phrase in my field notes. Nevertheless I have thought it appropriate to keep to the rule of providing a gloss for every form; there is nothing more frustrating than being unable to check a point in the phonological description of a language merely because the informant cannot identify the forms cited in the examples.

Most forms are cited in what I hope is a phonemic transcription; allophonic differences are rarely noted. But the transcription is not what Abercrombie (1953) calls simple phonemic. Whenever I have wished to draw attention to a particular phonetic feature of a language I have used the specific symbol which designates the appropriate phonetic categories. This is what Abercrombie calls a comparative (phonemic) transcription, since the reason for using a more exotic symbol is in order to compare the sounds of one language with those of another, or with some set of universal phonetic categories.

The present work has some relevance to the problem of the nature of universal phonetic categories. I would maintain that, as has been shown by Chomsky (1962), linguists must behave as if there were such categories; but that we are a long way from establishing a formal universal phonetic alphabet. Addicts of distinctive feature theory (Jakobson & Halle, 1956) will find plenty of new material here which challenges description in terms of the existing categories.

It is possible that some linguists may not know the value of all the symbols and diacritics used. I have tried to keep strictly within the usage prescribed by the International Phonetic Association, whose *Principles* (1949) provide a good guide. I must admit that, like many practising phoneticians, I would like to modify some I.P.A. statements quite considerably. Nevertheless I feel that this is no justification for setting up a personal set of symbols for the representation of sounds that can be adequately handled in I.P.A. terms. Only when there is no official I.P.A. symbol available for the representation of a sound have I felt it legitimate to coin a new symbol or borrow one from some other alphabet. Several other sets of symbols have been suggested by individuals (e.g. Bloch and Trager, 1942; Voegelin, 1959), but none of them is as internationally well known or as comprehensive as the I.P.A. alphabet. The only possible alternative for this book would be the 'Africa' alphabet of the International African Institute. This set of symbols is very similar to a section of the I.P.A. alphabet (it was, indeed, devised in cooperation with prominent members of the I.P.A.) except that instead of I.P.A. j (for the sound at the beginning of the English word 'yes') the symbol y is used; and instead of I.P.A. ɸ and β, the Africa alphabet symbolizes the bilabial fricatives by a long-tailed 'f' (ƒ) and ʋ. In this book the former symbol is not used; and ʋ is used to designate a bilabial approximant (in contrast with the labiovelar approximant symbolized by w). The only other point concerning symbols that need be noted is that the white spaces which occur in some phrases correspond to local orthographic divisions into 'words'. They do not necessarily have any phonetic value, and are put in simply for the convenience of the reader.

Adherence to the principles of the I.P.A. has resulted in transcriptions with a number of forms which differ from the local orthographies. Those who know some of the vernacular literature may find this inconvenient. But I hope their loss will be counterbalanced by the corresponding gain in consistency and universality which should help other linguists.

It is also inevitable that some readers will be unfamiliar with many of the instrumental techniques that were used to provide the illustrations. While I was working at University College, Ibadan, I was able to use: cine-radiology recording and analysing equipment, high-quality tape recorders and tape repeaters, a sound spectrograph, a pitch meter combined with an amplitude display unit, breath pressure and flow transducers, a high-speed ink writer, a palatograph, and a system for photographing the lips. The phonetic laboratories at the University of Ghana and the University of Dakar were also well equipped. At the moment there is no comprehensive handbook explaining the use of equipment of this kind; but some help may be gained from a number of recent works (Joos, 1948; Fischer-Jorgensen, 1958; Fant, 1958; Peterson, 1957; Ladefoged, 1962a). Since this book is not concerned with experimental phonetics as such, it is not appropriate to give more than a few additional notes and references here.

The cine-radiology research was conducted with the assistance of Dr. W. P. Cockshott and his staff. It involved the use of a Watson 1000 mA constant potential generator with electronic switching and a Machlett Superdynamax 150 X-ray tube with a 1 mm focus, in conjunction with a 7 in. Siemens image intensifier which was recorded on a 35 mm Arriflex camera at rates of between 30 and 40 frames per second. The camera shutter when fully opened triggered off the exposure which was delivered as a pulse of $1\frac{1}{2}$ to 2 milliseconds. The kilovoltage was between 60 and 70, and the milliamperage around 75 to 90. Through the use of the pulsing technique radiation was received only when the shutter was fully open and the afterglow of the phosphor was used effectively; in this way a low dosage was achieved. The maximum measured radiation at skin level was 0·6 mR per frame or 1·26 R during filming for one minute. The Double X Eastman film was developed in a Hadland continuous automatic processor, using May and Baker Exprol developer and Perfix fixer. It was analysed and tracings of individual frames were made with the aid of a Tage Arno multi-speed projector. Impressions of the oral cavity of each subject were made with Zelex dental impression material; sections of these impressions enabled accurate measurements to be made of the inferior aspect of the hard palate, which was often difficult to trace from the individual X-ray frames.

Two kinds of tape recorders were used: a battery-operated Kudelski Nagra in the field, and Ferrographs in the laboratory. All the tape recorders were constantly checked and serviced so that the signal to noise ratio was always better than 40 db, and the frequency response was always better than $\pm$ 2 db from 75 cps to 9,000 cps.

The tape repeater unit which was used was devised by J. Anthony (University of Edinburgh). It is probably the most useful single device available for the detailed study of the sounds of an utterance. This unit involves the use of a loop of tape in conjunction with an ordinary tape recorder which can reproduce the original tape recording. Instead of the original recording being reproduced over a loudspeaker in the usual way, the two units are connected together so that the original recording is first copied onto the loop of tape, and then reproduced from there. Before a piece of tape in this loop passes the recording head, it goes past an erasing head which deletes whatever has been recorded previously. Consequently, as the loop of tape goes round and round, new material is being constantly recorded on it in place of the old material which had been recorded a moment before; at any given instant the loop of tape contains a copy of the part of the original recording which has just been reproduced. When the operator wishes to hear a particular phrase repeated, he simply operates the stop switch on his tape recorder. The interconnexion of the tape recorder and the repeater unit has been arranged so that operating this switch not only stops the original recording but also cuts out the erasing head on the tape loop; but the loop is left running so that the part of the recording that has been copied onto it is reproduced over and over again. The joy of this device is its simplicity. The single switch can be manipulated with one hand without looking, so that the linguist can really concentrate on listening without having to worry about the mechanics of the operation.

The sound spectrograph was a Kay Electric Sona-Graph modified and used as described in Ladefoged (1962b). This device produces a visible display of the component frequencies of each sound. Fant (1960) has shown how many of the acoustic features of speech may be correlated with articulatory processes. Others (Moll, 1960; Truby, 1959) have also described the use of cine-radiology in conjunction with acoustic analyses. The data provided here may be of general interest in that they make it possible to correlate a number of less usual articulations with the corresponding acoustic features.

The pitch meter was built in the Speech Transmission Laboratory, R.I.T., Stockholm, by courtesy of Dr Gunnar Fant; it was a copy of the basic device discussed by Risberg (1962), which produces a voltage corresponding to the fundamental frequency. The amplitude display unit was similar to the Kay Electric model, but with a choice of time constants; it produced a voltage which was proportional to the logarithm of the intensity of the speech waveform. Thus these two devices enabled studies to be made of (roughly) the pitch and the loudness of the sounds on a recording.

The pressure of the air in the mouth and pharynx and the subglottal pressure were recorded with the aid of polyethylene tubes and RCA movable anode transducers as described in Ladefoged (1962c). The rate of flow of air out of the mouth was estimated from recordings of the pressure drop across a fine stainless steel mesh in the outlet of a face mask; this method of transducing airflow should be regarded as qualitative rather than quantitative.

The outputs of both the pressure and flow transducers, and the fundamental frequency and intensity devices were recorded on a three-channel Mingograph direct ink-writing oscilloscope. This recording system uses galvanometers supporting jets of ink; two of the channels had a flat frequency response up to 1,000 cps.

The use of palatography has been described in Ladefoged (1957). The basic technique involves spraying a dark powder on the upper surfaces of the informant's mouth, and then (usually) asking him to say a single word in which there is only one consonant contact between the tongue and the upper teeth or palate. When the tongue touches the roof of the mouth it wipes off some of the powder. The contact areas can be studied and photographed with the aid of a system of lights and mirrors; the resultant record is known as a palatogram. If the subject sticks his tongue out immediately after saying the word which is being examined it is also possible to see (and photograph) the parts of the tongue which have touched the powder on the roof of the mouth. This kind of record is called a linguagram. The tongue does not have the same shape when it is protruded as it does during the articulation, so linguagrams cannot be interpreted too precisely. But, as has been shown in Ladefoged (1957), palatograms of articulations in the front of the mouth can be interpreted with a fair degree of precision as long as there is a dental impression of the roof of the mouth available, so that the contours and sagittal section are accurately known.

The system of photographing the lips used the same camera and lighting as in the palatography device. But instead of a mirror in the mouth reflecting the upper teeth and the palate, another mirror was placed at an angle by the side of the face so that it was possible to photograph full-face and side-view simultaneously. When photographing the lip position during the sound in the middle of a word the appropriate timing was achieved by trial and error, checked by recording each utterance with a microphone placed near the camera, and then observing the position of the shutter click on a spectrogram of the recorded utterance.

The use of all these techniques furnished a great deal of information. I do not know of any previous attempt to use data provided by palatograms, linguagrams, casts of the mouth, photographs of the lips and spectrograms all of the same utterance, supplemented by tracings of cine-radiology films and pressure and flow recordings of similar utterances of the same word. But although instrumental techniques were used extensively in the course of this study, I would like to stress that an equally important part of the work consisted of simply observing and imitating informants. As a result of an experimental approach I was often able to select the most appropriate out of a number of conflicting hypotheses about the way in which a particular sound was made; and occasionally instrumental data revealed a new articulatory possibility which I had not thought of before. Nevertheless I am still sure that 'instrumental phonetics is, strictly speaking, not phonetics at all. It is only a help . . . The final arbiter in all phonetic questions is the trained ear of the practical phonetician.' (Sweet, 1911). For those of us who are not as skilled as Sweet, instrumental phonetics may be a very powerful aid and of great use in providing objective records on the basis of which we may verify or amend our subjective impressions. But even the most extensive array of instruments can never be a substitute for the linguist's accurate observation and imitation of an informant.

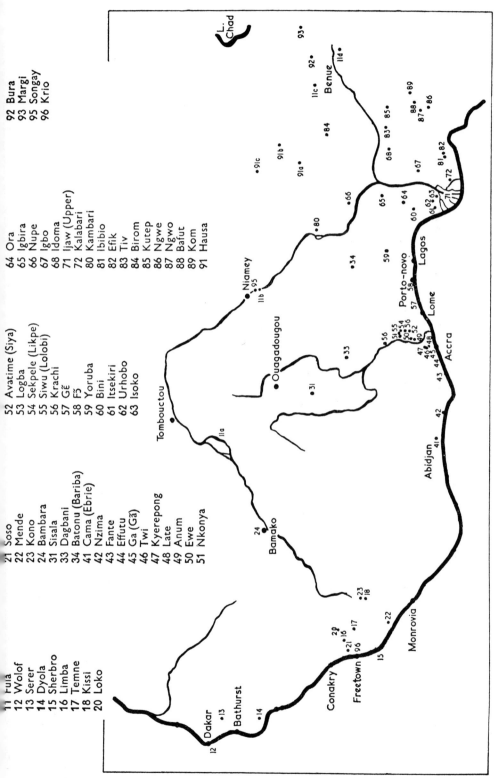

FIG. 1. Languages and hometowns of the principal informants listed in Appendix A.

11 Fula
12 Wolof
13 Serer
14 Dyola
15 Sherbro
16 Limba
17 Temne
18 Kissi
20 Loko

21 Soso
22 Mende
23 Kono
24 Bambara
31 Sisala
33 Dagbani
34 Batonu (Bariba)
41 Cama (Ebrie)
42 Nzima
43 Fante
44 Effutu
45 Ga (Gã)
46 Twi
47 Kyerepong
48 Late
49 Anum
50 Ewe
51 Nkonya

52 Avatime (Siya)
53 Logba
54 Sekpele (Likpe)
55 Siwu (Lolobi)
56 Krachi
57 Gẽ
58 F5
59 Yoruba
60 Bini
61 Itsekiri
62 Urhobo
63 Isoko

64 Ora
65 Igbira
66 Nupe
67 Igbo
68 Idoma
71 Ijaw (Upper)
72 Kalabari
80 Kambari
81 Ibibio
82 Efik
83 Tiv
84 Birom
85 Kutep
86 Ngwe
87 Ngwo
88 Bafut
89 Kom
91 Hausa

92 Bura
93 Margi
95 Songay
96 Krio

TABLE 1. Consonants occurring in West African languages. All the phonetic items shown contrast phonemically in at least one language with each adjacent item (irrespective of blank spaces), except for those items separated by heavy lines.

(a) Position of the soft palate	(b) Relation between the articulators	(c) Name summarizing (a) and (b)	(d) Position of unobstructed oral passage	(e) Action of glottis	labial alveolar	bilabial	labio-dental	dental	alveolar	post-alveolar	pre-palatal	labialised pre-palatal	palatal	velar	labialised velar	labial velar	labial palatal	uvular
(A) Upper articulator					lip & ridge	lip	teeth	teeth ridge	back of teeth ridge				hard palate	soft palate		and lip	palate & lip	uvula
(B) Lower articulator					lip & tip	lip	tip	tip or blade	tip	blade	front			back		and lip	front & lip	back
(C) Secondary articulators												lips			lips			
(D) Number on Figure 2					1 & 2	1	2	3	4	5	6	(1) 6	7	8	(1) 8	1 & 8	1 & 7	9
raised	complete closure	stop	(none)	voiceless ejective										k'				
				voiceless aspirated		ph			th				ch	kh				
				voiceless	p͡t	p		t̪	t				c	k	kʷ	k͡p		q
				voiced implosive		ɓ			ɗ									
				breathy voiced		bɦ			dɦ					gɦ				
				voiced		b		d̪	d	ɖ			ɟ	g	gʷ	g͡b		
				voiced laryngealised	ʔb͡d	ʔb			ʔd				ʔɟ					
	complete closure plus slow release	affricate	(none)	voiceless					ts	tʃ	tɕ	tɥ						
				voiced					dz	dʒ	dʑ	dɥ						
lowered	complete closure	nasal	(none)	voiced	m͡n	m			n				ɲ	ŋ	ŋʷ	ŋ͡m		
raised (unless contrary stated)	close approximation	fricative	central	voiceless		ɸ	f	θ	s	ʃ	ɕ	ɕɥ	ç	x		ʍ		X
				voiceless ejective					s'									
				voiced		β	v		z	ʒ	ʑ	ʑɥ	ʝ	ɣ		w		
			lateral	voiceless					ɬ				ʎ̥					
				voiced					ɮ									
	open approximation	approximant	lateral	voiced					l				ʎ					
			central	voiceless						ɹ̥						ʍ	ɥ̊	
				voiced		ʋ					ɹ		j	ɰ		w	ɥ	
				voiced laryngealised									ʔj			ʔw		
	(see Chapter 7) trill/tap		(none)	voiced					r/ɾ									
	flap		(none)	voiced			ⱱ		ɾ									

1

CONSONANT CONTRASTS

The consonants used in West African languages are very diverse. Before we examine the phonetic exponents of some of them in detail, it might be useful to consider the complete range that occurs in the languages discussed in this book. Table 1 gives the symbols for the contrasting phonetic elements used in the consonant systems that I had an opportunity to investigate. This display should enable the reader to get a rough idea of the value of the consonant symbols. Further assistance in this respect may be obtained from Figure 2, which shows the approximate parts of the vocal organs designated by the terms in Table 1. The numbered arrows in Figure 2 refer to the numbered categories in Table 1. None of the labels for the categories should be considered as indicating a precise description; they will all have to be qualified later. But we may note here that the terms labial alveolar, labial palatal and labial velar are used for items in which there are two equal places of articulation. (As opposed to labiodental, which denotes only one place of articulation, and labialized velar which denotes two articulations, one of which is secondary, having a lesser degree of stricture than the other.)

The main purpose of Table 1 is to demonstrate the multiplicity of phonetic contrasts which the investigator of West African languages must be able to handle. The number of descriptive categories could be reduced by means of a distinctive feature approach (Jakobson & Halle, 1956) or the abstraction of series generating components as suggested by Voegelin (1956). But neither of these possibilities reduces the number of events that the fieldworker must be prepared to recognize.

Many West African languages, including most of the Kwa group, can be considered to have no consonant clusters. The decision as to whether to regard the members of a particular sequence of consonants as single phonemic units or as clusters is, of course, often arbitrary. I have tried to include in Table 1 all the contrasting consonants in at least those languages that have a simple CV structure. Thus there is a column 'labialized velar' and another 'labialized post-alveolar' because in several languages distinctive labialization applies only to velars, or only to post-alveolars; and it is therefore convenient to regard these sounds as unit phonemes. But there is no column 'labialized alveolar' because all the languages which had sounds of this kind also had several other labialized sounds. Thus Birom has p, pw; b, bw; t, tw; d, dw; c, cw; ɟ, ɟw; k, kw; g, gw. The second of each of these pairs of sounds is usually regarded as a cluster (or the labialization may be regarded as a prosodic feature of a larger unit). Similar considerations apply to palatalized sounds. Welmers (1950) showed that in two related languages of the Senufo group it was convenient to consider one of them (Senadi) as having a series of palatal consonants, and the other (Sup'ide) as having several palatalized consonants, which he interpreted as clusters. Clusters with r or l, one or other of which can occur between each single consonant and a vowel in a single syllable in Ewe (Ansre 1961), are also omitted. Table 1 may be regarded as an array of contrasting phonetic elements; the symbols listed form the minimum selection from the universal phonetic alphabet that is needed for the phonetic categorization of the segmental phonemes of these West African languages; and if some of the finer points considered in subsequent chapters were taken into account, then the array presented here would have to be greatly extended. Thus

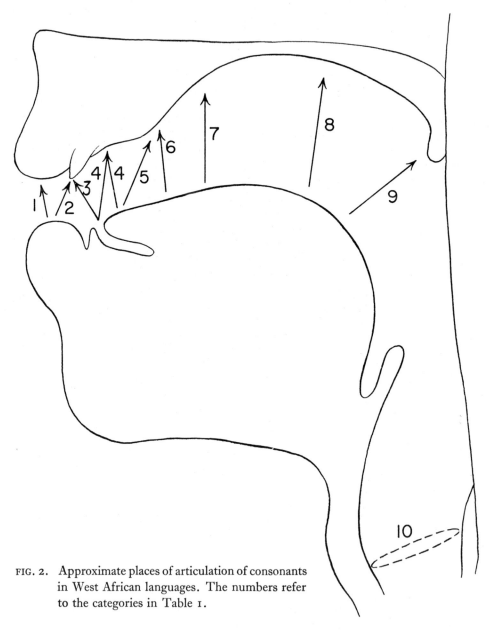

FIG. 2. Approximate places of articulation of consonants
in West African languages. The numbers refer
to the categories in Table 1.

obviously the column headed post-alveolar could be subdivided; the articulators involved
in ʒ and the different forms of ɹ are not at all alike. But it would also be possible to
reduce the number of categories by following the suggestions of Jakobson and Halle
(1956). Thus there are languages where p contrasts with k͡p, and others where p contrasts
with p͡t; but there are no languages where k͡p contrasts with p͡t. So these two sounds
could be regarded as belonging to the same phonetic category. Intuitively, this seems a
reasonable solution in this case; but it is difficult to know where to stop combining items
into a single category. No West African language happens to have k͡p and q in contrast;

2

should these two items be placed together? And what about ɗ and dh, which do not happen to contrast in any of these languages (as opposed to ɓ and bh, which contrast in some forms of Igbo)? Since the object of this book is to give a detailed account of the phonetic events which can be correlated with the linguistic contrasts in West African languages, I have thought it better to show too many rather than too few phonetic elements in Table 1.

Many of the pairs of sounds represented by adjacent symbols form contrasts which are hard for European observers to distinguish. Nevertheless nearly every possible pair represents a contrast which is linguistically significant somewhere in West Africa. Except in the case of those items which are separated by a heavy line, each item contrasts in at least one language with each item above and below it in the column, and with each of the items on either side of it in the row (irrespective of whether there are blank spaces between the adjacent items). Examples of the use of the contrasts between most of the adjacent items and other similar phonetic elements are given in Table 2. This table provides a quick source of some of the less well known phonemic oppositions. Further examples of these and many other oppositions can be found in Appendix B.

TABLE 2. Examples of contrasts between some of the items in Table 1, and some other similar phonetic elements.

k'–k	Hausa:	wáːk'àː	song	kàːká:	grandparent		
ts–tʃ	Nupe:	tsa	choose	tʃa	begin		
dz–dʒ		dzami	bridle	dʒama	assembly		
tɕ–dʑ–ɕ	Gã:	étɕò	it is ablaze	èdʑà	it is right	éɕá	sin
tɕʷ–dʑʷ–ɕʷ		etɕʷa	he struck	èdʑʷa	it's broken	éɕʷá	it is scattered
tʃ–tɥ–c –ch	Cama:	tʃa	stamp	tɥɛ	learn	cà chà	destroy pound
ph–th–kh	Igbo: (Owerri)	àphé	sharpening	àthá	blaming	ákhá	long
p–t–k		ápè	pressing	àtá	chewing	áká	hand
b–d–g		mbà	another town	ádà	crab	àgá	going
bɦ–dɦ–gɦ		mbɦá	boast	ádɦà	falling	àgɦá	being useful
ɓ–ɓɦ		àɓà	power	àɓɦà	jaw		
p͡t–p–t b–d	Bura:	p͡tá	*animal	pàkà bàrà	search want	tá dàwà	cook enemy
ʔbd–ʔb–ʔd		ʔbdà	chew	ʔbáɬà	dance	ʔdà	eat meat
t–d̪	Isoko:	òtú	louse	úd̪ù	farm		
t–d–n		òtú	gang	úd̪ù	chest	òna	skill
f–θ–t	Sherbro:	faká	village	θàm	grandmother	tàmbàsé	sign
d–d̪–ɟ	Logba:	ɔ́dɛɛ	he scolds	ɔ̀d̪è	*clan	ɔ̀ɟándɛ́	gap in teeth
c–k	Urhobo:	écá	coming	ɔ́kà	maize		

3

TABLE 2. Examples of contrasts between some of the items in Table 1—continued

k͡p–kʷ		ɔ̀k͡pè	ugly	ɔ́kʷa	he packs		
ɟ–g		oɟa	soap	ɔ̀gà	illness		
g͡b–gʷ		og͡ba	fence	ɔ́gʷà	*yam		
ɓ–b–g͡b	Kala-	aɓa	*girl	àbà	*fish	àg͡bà	offering
p–ɗ–d	bari:	àpà	fool	áɗa	dad	àda	eldest daughter
k–q–χ	Serer:	kor	man	qos	leg	χol	clean
m–n–ŋ	Idoma:	áma	bell	ànà	*fruit	ɔŋáɟi	rainbow
ɲ–ŋʷ–ŋ͡m		áɲà	quick temper	àŋʷà	divining pods	aŋ͡màa	body marks
ɸ–f	Ewe:			éɸá	he polished	éfá	it was cold
w–β–v		éwɔ̀	he made	ɛ̀βɛ̀	Ewe language	ɛ̀vɛ̀	two
s–s'–ʃ	Hausa:	wàːsáː	play	s'áːs'àː	rust	ʃáːʃàːʃáː	fool
s–ʃ	Urhobo	ɔ̀sè	feather	èʃà	grey hair		
z–ʒ		òze	baisin	oʒa	suffering		
ʃ–ç–h	Bura:	ʃá	lost	çál	guts	hala	get old
x–h	Ora:	ɔ̀xà	story	ɔ̀hà	wife		
ʛ–ɬ–ʎ	Bura:	ʛábʷá	beat	ɬá	cow	ʎálá	cucumber
l ʎ		là	build			wúʎá	neck
l–ɹ–ɻ	Bini:	álázi	*monkey	aɹá	caterpillar	áɹába	rubber
ɻ̣				àɻà	burial ceremony		
l–ɹ–ɾ	Isoko:	òlá	jump	òɹá	flight	òɾá	yours
ʍ–ɥ–h	Birom:	ʍà	wife	ɥégèlèk	yesterday	hòm	cheek
jˡ–j–ʔj	Margi:	jˡàjˡàʔdò	picked up	jà	give birth	ʔjà	thigh
wˡ–w–ʔw		wˡá	reach inside	káwà	sorry	ʔwáʔwí	adornment
j–ɥ–w	Idoma:	ɔjá	width	ɔɥá	moon	ɔ́wá	redness
ɾ–ɽ	Hausa:	báɾàː	begging	báɽàː	servant		

2

AIRSTREAM MECHANISMS AND DOUBLE ARTICULATIONS

We may begin our more detailed examination of the phonetic nature of some of the consonant elements in West African languages with a discussion of the airstream mechanisms which may be involved. As Catford (1939) and Pike (1943) have shown, there are three principal methods of moving air to form speech sounds: (1) the PULMONIC airstream mechanism, in which the air in the lungs is moved (usually outwards) by the action of the respiratory muscles; (2) the GLOTTALIC airstream mechanism, in which the air in the pharynx is moved inwards or outwards by the movement upwards or downwards of the closed glottis; (3) the VELARIC airstream mechanism, in which the air in the mouth is moved (usually inwards) by the movement (usually backwards) of the point of contact between the raised back of the tongue and the roof of the mouth. Most sounds are made with a pulmonic airstream mechanism; but all three mechanisms are used in some West African languages. In some sounds two mechanisms are used simultaneously; and occasionally all three mechanisms function in the production of a single sound.

The pulmonic airstream mechanism will not be discussed here; it appears to be used in essentially the same way as in Indo-European and all other languages. The glottalic airstream mechanism is not used in the well known European languages (except as a stylistic variant); but in West African languages it is involved in the production of two kinds of sounds: ejectives, in which the closed glottis moves rapidly upwards and compresses the air behind an articulatory closure; and implosives, in which there is a downward movement of the vibrating glottis, which tends to cause a lowering of the pressure behind the oral closure.

In the sixty-one languages investigated, ejectives occurred only in Hausa, which has c' k' kʷ' s' (with some dialectal variation: Katsina Hausa has ts' instead of s'). The stops in this series were accompanied by an interval of 50–60 msec between the release of the closure and the start of the vowel. This is slightly longer than the period of aspiration following the corresponding stops c k kʷ, which usually lasted about 35–45 msec. (Hausa stops are sometimes said to be unaspirated; but this was not true in the case of my informants.) The increase in the voiceless period may be associated with the time required for the glottis to return to the position necessary for normal voicing.

In the ejective fricative s' the up-and-down movement of the glottis can be correlated with a movement of the point of articulation which causes a variation in the pole frequency associated with the fricative noise. Plate 1a is a reproduction of a spectrogram of the Hausa phrase s'úns'àːjénè 'they are birds'. The variations in frequency can be seen very clearly in the medial s', and to a slightly lesser extent in the initial one. The upward movement of the glottis is achieved by an upward movement of the hyoid bone accompanied by the contraction of the thyrohyoid muscle which links the hyoid bone and the thyroid cartilage. The hyoid bone is a small horseshoe-shaped bone below the base of the mandible; it forms the foundation for many of the muscles of the tongue. If it is moved upwards the position of the body of the tongue will be altered so that the articulatory constriction is moved forwards. This alteration causes the variation in the fricative noise which we have noted. Plate 1a also shows that there is an interval of about

5

40 msec between the end of the recorded fricative sound and the release of the glottal stop, and a further short interval before the vocal cords start vibrating regularly.

The connexion between the hyoid bone and the larynx, discussed above, often has important consequences which are not usually noted in phonetic studies. The action we have been considering is one in which the movement of the larynx affects the articulation. Later on, when considering the interaction of tone and vowel quality, we will have occasion to refer to a case where the link between the hyoid bone and the larynx results in an articulation affecting a laryngeal function.

Sounds with an ingressive glottalic airstream mechanism were found in some West African languages; but they were not as common as the literature implies. Published phonetic descriptions indicate that there is often a confusion between sounds such as ɓ in Igbo and Kalabari, in which one of the essential components is a downward movement of the glottis, and ʔb in the West Atlantic (Fula, Serer, Wolof) and Chadic (Hausa, Bura, Margi) groups of languages. In the latter case there may be a downward movement of the glottis, but the essential component is a particular mode of vibration of the vocal cords; accordingly these sounds are discussed in the next chapter on 'Phonation Types'.

In the West African languages in which glottalic ingressive sounds were recorded, it was found that (as in the case of nearly all the other languages in the world) the downward movement of the glottis occurred while the vocal cords were vibrating. The vibrations were maintained by a small amount of lung air, which was under pressure from the respiratory system, being allowed to pass between the descending vocal cords. These sounds, therefore, involved a glottalic ingressive airstream mechanism, combined with a pulmonic egressive one. The downward movement of the vibrating glottis tended to lower the pressure of the air in the mouth; but this was usually more than offset by the increase in pressure due to the outgoing lung air. These sounds were seldom ingressive in the sense that on the release of the articulatory closure air flowed into the mouth. There are in fact not two clearly defined entities b and ɓ, but a gradient between one form of voiced plosive and what we may call a true implosive. A voiced plosive may have no downward movement of the glottis at all; but quite often (even in an ordinary English b) there is a lowering of the larynx which enlarges the oral cavity and thus allows a greater flow of air to keep the vocal cords vibrating (Hudgins & Stetson, 1935). Few people would call such a sound an implosive; but the mechanism is essentially the same as in Kalabari ɓ, differing only in the rate and degree of larynx movement.

The bilabial implosive ɓ which occurs in Igbo (and is spelt 'gb' in the local orthography) was investigated in the pronunciation of speakers of three different dialects: Owerri Province, Onitsha and Umuahia. All three dialects were found to be essentially the same from this point of view. Recordings of the pressure of the air in the mouth showed that when reading short sentences and word lists the pressure was seldom (8 per cent of the time) negative, and even then by not more than 2 cm aq; usually (74 per cent of the time) it was between 0·25 and 0·75 of the pressure during a pulmonic egressive b in comparable circumstances; but occasionally (about 4 per cent of the time) it was as great as in the formation of a similar b.

These figures indicated that Igbo ɓ might be distinguished from b by some feature other than the lowering of the larynx, since this obviously did not occur to any great extent on all occasions. Accordingly the articulatory movements of all three speakers were investigated with the aid of cine-radiology. The results indicated that ɓ was velarized as

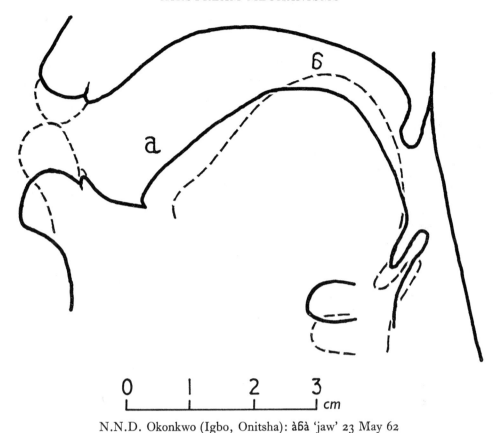

N.N.D. Okonkwo (Igbo, Onitsha): àɓà 'jaw' 23 May 62

FIG. 3. Tracings from frames of a cine-radiology film illustrating implosive ɓ in Igbo.

well as usually involving lowering of the glottis. Tracings from frames taken during a typical utterance are shown in Figure 3. It may be seen that there is a considerable raising of the back of the tongue (which did not occur in words containing b) and a downward movement of the hyoid bone, which is attached to the larynx; the glottis itself was not recorded on this film, as the size of the image did not permit viewing the lips and glottis simultaneously. But we may infer a little about the extent of its movement from observation of the movement of the hyoid bone. This movement was almost certainly due to a downward pull from the larynx, which is itself being pulled down, probably by the sternothyroid muscle. The acoustic effect of these movements will be considered later (page 13). In this particular utterance a negative pressure of between –1 and –2 cm aq was recorded during the bilabial closure.

The Owerri town dialect of Igbo has implosive ɗ in addition to ɓ (Ward, 1936). Similar ɓ and ɗ sounds also occur in Kalabari; but in this language the ɓ is always implosive and not velarized. Since Kalabari is one of the few languages that have ɓ and g͡b in contrast, it is easier to return to this point after discussing the production of labial velars.

A velaric airstream mechanism of the kind that produces the clicks which are common in Zulu, Xhosa and other South African languages does not occur by itself in any West

AIRSTREAM MECHANISMS	PRESSURE RECORDS	EXAMPLES

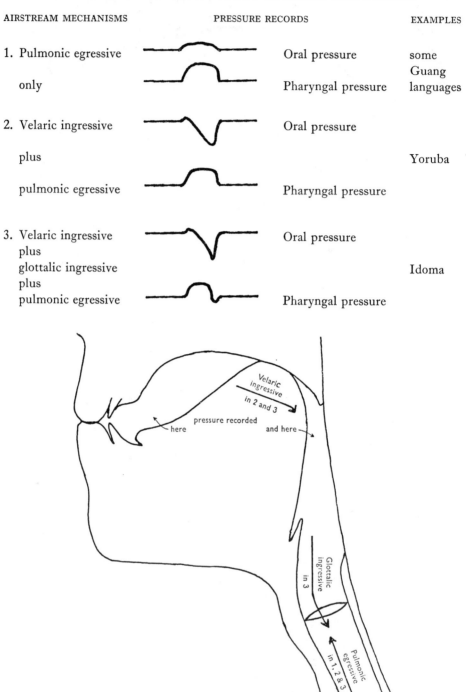

1. Pulmonic egressive

   only

   Oral pressure

   Pharyngal pressure

   some Guang languages

2. Velaric ingressive

   plus

   pulmonic egressive

   Oral pressure

   Pharyngal pressure

   Yoruba

3. Velaric ingressive
   plus
   glottalic ingressive
   plus
   pulmonic egressive

   Oral pressure

   Pharyngal pressure

   Idoma

FIG. 4. Types of airstream mechanisms involved in forming k͡p. The numbers by the arrows refer to the types listed above.

8

African language. A velaric mechanism is, however, used in conjunction with the pulmonic mechanism in the formation of many of the sounds written k͡p and g͡b. These sounds are formed in at least three different ways, which are summarized in Figure 4. The first type occurs in many Guang languages (Late, Anum). It consists of simply the simultaneous articulation of k and p or g and b, superimposed on a pulmonic airstream. The second type, which is found in Yoruba, Ibibio and many other languages, is more complicated. After the two closures have been made, there is a downward movement of the jaw, and a backward movement of the point of contact of the back of the tongue and the soft palate; these movements cause a lowering of the pressure in the mouth. Thus from the point of view of the release of the closure at the lips, there is an ingressive velaric airstream. But there is still a high pressure behind the velar closure owing to the outgoing air from the lungs (the pulmonic egressive airstream mechanism). Consequently when both closures are released the air flows into the mouth from two directions. This combination of a velaric and pulmonic airstream mechanism has been described very accurately by Siertsema (1958a) who concluded that Yoruba k͡p 'is implosive at the lips, "explosive" at the back'. Welmers (1950) also states that in the case of Senadi: 'During the stop, there is noticeable suction in the oral cavity, with a resultant "pop" at the moment of release.'

I have been able to substantiate the description of this kind of k͡p by means of instrumental techniques. A small tube between the lips was used in determining the pressure in the mouth between the two closures: and another tube passed through the nose into the pharynx to a point just above the vocal cords formed part of the system for measuring the pressure behind the velar closure. Plate 2a shows the pressures recorded when an Itsekiri speaker said ìpèré k͡pòrò búg͡bá 'a big nail and a calabash'. It may be seen that in the p in the first word the pressure in the mouth is the same as that in the pharynx; whereas in the k͡p in the next word the mouth pressure goes down (because of the lowering of the jaw and the backward movement of the tongue), while the pharyngal pressure stays up. Similarly in the b the two are the same (but lower than in the p, as the pressure of the air above the vocal cords must be less than the pressure in the lungs, if the vocal cords are to continue vibrating); but in g͡b the two pressures go in opposite directions. In both the k͡p and the g͡b the mouth pressure goes up sharply before the pharyngal pressure goes down, indicating that the velar closure is released before the labial closure.

In the third type of k͡p, which is found in Idoma and sometimes in Bini, all three airstream mechanisms are involved. After the two closures have been made there is a backward movement of the tongue as in the Itsekiri k͡p; and during the latter part of the sound there is also a downward movement of the vibrating glottis. Thus, despite the orthography, these sounds are in part voiced. This kind of k͡p, in which velaric ingressive, glottalic ingressive and pulmonic egressive airstream mechanisms are employed simultaneously, occurs frequently in Idoma and Isoko: but in Bini it may be in free variation with one of the other possibilities already described. Plate 3a shows the pressure records of the Bini phrase ɔ́fíɛ̀ ipàpà ɔ́k͡pá ʋè ík͡pák͡pa g͡bèɹé fua 'he cut a slice off his skin'. In this one utterance all three kinds of k͡p occur. The third type described above occurs first; the k͡p marked (1) has a velaric ingressive airstream in the first part of the sound (as is shown by the mouth pressure decreasing while the pharynx pressure remains at its highest level); but in the latter part of the sound there must be a voiced ingressive glottalic mechanism, since the pharynx pressure drops below normal, and at the same time vibrations due to voicing can be clearly seen. The drop in pressure

9

can be caused only by a lowering of the vibrating glottis. (An ingressive pulmonic airstream mechanism is not known to occur interspersed with the normal pulmonic egressive airstream in any language, probably because the respiratory muscles cannot easily reverse the direction of air flow within the duration of a single consonant.) The k͡p marked (2) is of the type already illustrated in the Itsekiri example, and need not be discussed further. The k͡p marked (3) is an example of the first type listed in Figure 4. The mouth pressure is varying very slightly above normal; so, despite the fact that there is a velar closure (as is shown by there being a difference between the oral and the pharyngal pressure), there is no movement causing a velaric ingressive airstream. This k͡p is simply the superimposition of two simultaneous stop closures on a pulmonic airstream. The single example of g͡b in this utterance is the voiced equivalent of the k͡p marked (2): the oral pressure record shows that there is a velaric ingressive airstream; and the pharyngal pressure record shows that there is a pulmonic airstream; the ripple due to the vibrations of the vocal cords can be seen on both records.

My data do not allow me to say whether the differences between the first and second k͡p in the phrase in Plate 3a are due to free variation in Bini, or whether they are determined by the phonetic environment. But we may note that in seven utterances of this phrase the third k͡p always had a slight positive mouth pressure; and twice out of the seven times the mouth pressure and pharyngal pressure were exactly the same, indicating that there was not a complete velar closure. It may well be that when Bini k͡p occurs as the second consonant in a word of this type, it is articulated less forcibly.

The g͡b which occurs in contrast with both b and ɓ in Kalabari is illustrated in Plate 2b. Unfortunately I was unable to record the pharyngal pressure in this utterance; but it is readily apparent that the g͡b involves a combination of the pulmonic and velaric airstream mechanisms of the kind that we have discussed already. The record shows that on this occasion the informant closed his lips very slightly before making the velar closure (since the mouth pressure goes up to begin with); then there is a period of about 30 msec during which the back of the tongue must have been pressed fairly steadily against the roof of the mouth (since the mean pressure is constant, and the pressure variations due to the vibrations of the vocal cords are much smaller than those recorded in the front part of the oral cavity in the other bilabial consonants in which there is no velar closure); this period is followed by the enlargement of the oral cavity, presumably due at least in part to the backward movement of the tongue, which took about 60 msec (during which the pressure decreased and became negative); and then the velar closure must have been completely released while the lips remained closed for a further 50 msec (since the pressure suddenly increases, and there are large voicing vibrations).

The formation of this sound may be compared with the production of the pulmonic egressive b in the first word, in which the mouth pressure increases almost throughout; and with the ɓ in the last word in which the glottis must have been descending, since the pressure starts slightly positive and then becomes slightly negative. The maintenance, or small increase, in the amplitude of the voicing vibrations in this is an indication that the glottis is vibrating in the same way as in the formation of other voiced sounds. This distinguishes this stop from the Hausa stops ʔb ʔd which will be considered in the next chapter; in these sounds the amplitude of the vibrations is considerably reduced, due to the short length of the glottis which is vibrating.

Table 3 summarizes the data I have on the occurrence of the different forms of k͡p in the sixty-one languages which were investigated.

TABLE 3. Distribution of the different types of k͡p and velarized p in the 61 languages

Type on Figure 4	Airstream mechanism	Phonation type	Articulation	Number of languages
—	pulmonic egressive	voiceless	velarized bilabial	4
1	pulmonic egressive	voiceless	labial velar	2
2	pulmonic egressive plus	voiceless	labial velar	23
3	velaric ingressive	part voiced	labial velar	8

The value of this table is strictly limited. Firstly it must be remembered that it is based on observation of very few informants for each language. Secondly the placing of a language in one category as opposed to another is often arbitrary; there is sometimes considerable variation in the production of different examples of k͡p within a language, a point that has already been made with regard to the examples of k͡p in Bini shown in Plate 3a. Thirdly the use of a limited number of categories necessarily obscures some of the facts. We have already noted that languages such as Idoma consistently use an ingressive glottalic airstream mechanism in conjunction with the ingressive velaric and egressive pulmonic mechanisms to produce a partly voiced k͡p. I am fairly sure that Kono and some of the other languages counted as having partly voiced k͡p also use the three mechanisms simultaneously. But as my data do not include oral and pharyngal pressure records for all these languages, I cannot state for certain when a glottalic ingressive mechanism is also involved.

Twenty-four of the thirty-seven languages with some form of k͡p also have the fully voiced counterpart g͡b. Only Kambari and Kalabari have the three consonants k͡p–g͡b–ɓ; and only Igbo has ɓ (but no g͡b) in contrast with k͡p. Two languages (Sherbro and Limba) have labial velars which are not counted in Table 3, since they have only the voiced forms. In both these languages the voiced labial velar may be said to be primarily opposed to the voiceless velar in the stop system: p t tʃ k; b d dʒ g͡b. Hockett (1955) suggested that there were no languages which had both k͡p and kʷ. In fact twelve of the thirty-seven languages which have k͡p also have kʷ; and in at least eight of these languages consonantal sequences involving w only occur with velar consonants, so that in these languages the labialized velars are unit phonemes on a par with the labial velars.

Hockett (1955) also indicates that the only known kind of coarticulation is that involving simultaneous labial and velar strictures. It is by far the most common kind of double articulation; but there are also languages which have labial alveolar stops such as p͡t and b͡d. In Gur languages such as Dagbani these sounds are allophones of k͡p and g͡b which occur before front vowels. But in the Chadic languages Bura and Margi the labial alveolar mode of articulation is clearly on an equal standing with all other points of articulation. A pulmonic airstream mechanism is used in the formation of p͡t which contrasts with p t k (but not with k͡p) in all contexts; and there are also the affricated forms p͡ts and p͡tʃ, the laryngealized form ʔb̰d̰, and the homorganic nasal m̄n̄.

A velaric airstream mechanism is also used in the production of nasals. Some languages, such as Idoma, have a nasal ŋm̄ which has a double articulation similar to the stops k͡p and g͡b. Plate 3b shows the pharyngal and oral pressures recorded during two Idoma words. The bilabial click component which accompanies the velar nasal can often be heard when the lip closure is released. It is not, however, very audible, and is thus quite different from the nasal clicks which occur in South African languages such as Xhosa, where the negative pressure produced during the articulation is very much greater.

We may conclude this chapter with a few remarks on the acoustic characteristics of the labial velar stops k͡p g͡b. There is a well known auditory resemblance between these sounds and the corresponding velarized labials pʷ bʷ. Westermann and Ward (1933) give examples of variation between k͡p g͡b and kʷ gʷ in dialect forms of some languages; and Abraham (1940) noted the tendency of some examples of k͡p in Tiv to have a labial-ized quality (he writes kpw before e in complementary distribution with k͡p before all other vowels).

Spectrographic data offer clear evidence of this labialized quality. Plate 4a illustrates two Yoruba words. In each of these the movement of the second formant which occurs on the release of the stop consonant is in the upward direction, as is normal for segments involving initial labial velar quality. The rate of change is slightly faster than it would be for a sequence of stop plus semivowel, but not quite as fast as occurs after a plain bilabial stop. An interesting effect observable in the first word is the tendency of the second formant to have two points of origin ('hub' 'locus'), one at about 1200 cps and the other at about 2200 cps; this effect could often be seen in the case of labial velars before high front vowels.

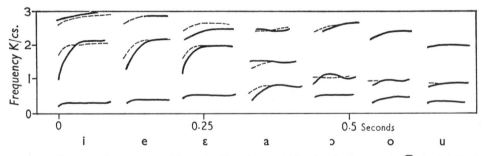

FIG. 5. Average formant positions in Yoruba syllables beginning with g͡b (solid line) and b (dashed line where different from g͡b) before each of the seven oral vowels.

The distinction between g͡b and b in Yoruba is illustrated by the diagram in Figure 5 which shows the first three formants for g͡b (solid line) and b (broken line when different from g͡b, otherwise solid line) before each of the seven oral vowels in Yoruba. The diagram represents the mean positions in a number of different utterances (at least two each by two informants). The data were averaged by eye on the basis of tracings from spectro-grams of whole words; only the appropriate parts of each word have been reproduced in this figure.

The acoustic differences between the Yoruba labial velar and labial stops are mainly in the second and third formants; only when the first formant is very high, as in a, is the velarity feature conveyed partly by a different transition of the first formant. In the case of all the other vowels the second formant has a larger transition when the initial consonant is g͡b. This might be expected on the grounds that when there is a velar closure as well as a bilabial closure there must be a greater movement to get to a vowel position; when there is just a bilabial closure the tongue can be already forming the vowel before the lips open. The required articulatory movement is less before back vowels; as can be seen the transitions are smaller in these circumstances. (I cannot account for the transitions in the case of the syllable ba; the second formant had a larger transition in this syllable than in the syllable g͡ba on a statistically significant number of occasions.)

The acoustic characteristics of implosive ɓ are in some ways similar to those of g͡b. Plate 4b illustrates the formant transitions that occur in four Igbo words. The glide after the release of the bilabial closure is partly due to the velarization which we have noted already; but it can be associated also with the return of the depressed larynx to the normal position. When the glottis is lowered and the pharyngal cavity enlarged there is a corresponding decrease in the frequency of the formants which, in this case, extends through the greater part of the whole syllable. This feature probably accounts for most of the particular phonetic quality associated with implosive consonants.

3

PHONATION TYPES

In many Indo-European languages the state of the glottis can be correlated with two or at the most three sets of linguistic contrasts: in English the vocal cords function in different ways in the formation of voiced and voiceless sounds, and (according to Lehiste, 1962) in still another way in the formation of h sounds. Most West African languages use at least these three different states of the glottis; out of the sixty-one languages investigated, only seventeen had no form of h sound.

Some West African languages also use distinctions which depend on the timing of these actions of the glottis. I have followed the usual tradition (well described in Abercrombie, forthcoming) of restricting the terms voiced/voiceless to describe the state of the glottis during an articulation, and the terms aspirated/unaspirated to describe the state of the glottis during and immediately after the release of an articulation. In this sense Cama is the only language in our sample that has a regular three-term system of stop consonants: voiced/voiceless-unaspirated/aspirated. Examples (suggested by J. M. Stewart) showing the contrast between b–p–ph; d–t–th; c–ch; g–k–kh have been recorded; some of them are given in Table 2 (pp.3–4).

This set of distinctions, which is common in the languages of S.E. Asia, is fairly rare in West Africa. But a number of languages use the aspirated/unaspirated contrast to supplement the distinction between sounds which have similar articulations. Thus in Soso p is aspirated, k͡p is voiceless unaspirated, and b is voiced (there is also a fully voiced g͡b which verges on the implosive). Many other languages supplement the distinction between p and k͡p in this way: Abraham, who is an acute if somewhat alinguistic observer, writes kb for k͡p in some of his earlier work on Idoma (Abraham, 1933); and, as can be seen in Plate 2a, Itsekiri p is more aspirated than k͡p. In other languages additional articulations, such as the different forms of t in Temne, are also partially distinguished by differences in aspiration (see page 19).

The Igbo dialect which is spoken in parts of Owerri Province (but not in the town of Owerri) uses the distinction between aspirated and unaspirated stops as well as the voiced/voiceless distinction. In this case there is a four-term system, which is given (Carnochan, 1948) as: voiced; voiceless unaspirated; (voiceless) aspirated; voiced aspirated. The first three terms are as in Cama; the distinguishing parameter is the time of onset of normal vocal cord vibration (voicing) in relation to the formation, hold and release of an articulatory closure. The fourth term, from a phonetic point of view, is really in a different category; the distinguishing parameter is the 'breathy voice' mode of vibration of the vocal cords that occurs after the release of the closure. In this state of the glottis, which may be symbolized by ɦ, the vocal cords are vibrating, but there is a comparatively high rate of flow of air out of the lungs (Catford, 1963; Ladefoged, 1963). High-speed motion pictures of similar sounds show that the vocal cords are vibrating but not closing completely (Uldall, 1958).

Plate 5 provides data illustrating the difference between Igbo p–ph and ɓ–ɓɦ). It may be seen that in the unaspirated p the lips close (about 50 msec before the vocal cords stop vibrating) and the pressure in the mouth remains constant; after the lips open there is only a short interval (about 25 msec) and a low rate of flow of air before the vocal cords start vibrating for the vowel. But in the aspirated ph the lips close (again shortly before the vocal cords stop vibrating) and the pressure in the mouth increases considerably during the closure; then the lips open and there is a very high rate of flow of air out of the mouth, and a slightly longer interval (about 55 msec) before the vocal cords are vibrating regularly. The lower part of the figure shows the very interesting contrast between the voiced implosive ɓ and the breathy voiced implosive ɓɦ. In the case of the first sound the spectrogram and the expanded waveform shows that normal voicing vibrations persist throughout the closure; the pressure record shows that the mean oral pressure is slightly negative in the middle of the articulatory closure but at the moment when the lips are about to come apart it is approximately zero; consequently there is no measurable flow of air in either direction when the closure is released. In the breathy voiced implosive ɓɦ the mean oral pressure is also slightly negative during part of the articulation; but after the release of the closure there is a high rate of flow of air, despite the fact that voicing vibrations persist. The spectrograms and expanded waveforms show that this kind of voicing produces little acoustic energy.

I am not sure, but it is possible that another state of the glottis may be involved in the formation of some final consonants in Wolof. The phonological situation is quite clear: in initial position there are contrasts which are phonetically p–b, t–d, g–k–q (or, diaphonically, χ instead of q); in final position there are the same number of similar sounds, but none of them are voiced. It is not difficult to make a phonemic statement about these data; we need recognize only the phonemes p b t d g k q. But the problem of the phonetic nature of the contrast in final position remains. Ward (1939a) states that the final stops in Wolof contrast in that some are exploded, whereas others are unreleased. There is some evidence (Sauvageot, personal communication) that this is still true in the outlying areas where the influence of French is not so powerful. But my own studies indicate that at least in the present-day spoken Wolof of Dakar the contrast is made in other ways. Plate 6 illustrates this point by spectrograms of two minimal pairs. In the first pair the t is clearly exploded at the end of both words; but the one explosion is very much more forceful than the other. This is apparently one of the rare cases where the ostensible major difference between two sounds is that one is fortis and the other is lenis. But even in this case my own impression is that the difference cannot be correlated with greater, as opposed to less, force being exerted by the respiratory system (or, indeed, any other part of the vocal apparatus), which would seem to be the proper domain of the fortis/lenis labels. On the basis of the acoustic data and my own hearing and imitation of the informants, it seems that the state of the glottis is the more likely physiological correlate. During the release of the stop in the first word the vocal cords are probably restricting the flow of air by being closer together as in the position for whisper; I do not know whether the narrowing follows a glottal closure which may have occurred during the stop. But irrespective of whether there is at any time a complete glottal closure or not, it would seem that the less forceful explosion can probably be correlated with the greater muscular tension (of the vocal cords) – a result that hardly agrees with the usual accounts of what is meant by lenis as opposed to fortis.

The lower part of Plate 6 illustrates the contrast between what is usually described in phonemic terms as χ as opposed to q. (There is also a velar stop contrasting with these two sounds in a word such as ɲak 'vaccine'.) Both final consonants in the words in the lower part of Plate 6 clearly have the same place and manner of articulation; they are both uvular fricatives. They differ only in the intensity of the fricative noise. It is at least possible that this difference, which is dependent on the rate of flow of air through the fricative stricture, may be due to a variation in the airflow associated with a whisper as opposed to a voiceless position of the glottis.

So far we have considered five states of the glottis which may have a linguistic function in West African languages: (1) voice, in which the vocal cords are vibrating as a whole; (2) voicelessness, where the vocal cords are apart as in normal expiration; (3) h, in which the opening between the cords is slightly less than in (2) (these last two are usually in complementary distribution and do not form a linguistic opposition in any languages that I know of); (4) whisper, as in some so-called lenis sounds, in which there is considerable narrowing, perhaps leaving a gap only posteriorly between the arytenoid cartilages; and (5) breathy voice, in which the vocal cords vibrate loosely without touching one another. A sixth and most important state is known as laryngealized voice, or following Catford (1963) 'creaky voice'. In this state of the glottis there is a great deal of tension in the intrinsic laryngeal musculature, and the vocal cords no longer vibrate as a whole. The ligamental and arytenoid parts of the vocal cords vibrate separately, sometimes almost exactly 180° out of phase with one another, one end opening as the other is closing. This produces an apparent increase, often an approximate doubling, of the rate of occurrence of glottal pulses. Laryngealized voicing often occurs during an implosive consonant. In the lower part of Plate 5, which has been discussed already, the points above the expanded wave form show the onset of each glottal pulse; it may be seen that in the middle of the consonant the frequency is almost doubled. The pitch of sounds with this sort of voicing is not, however, higher than that of the surrounding normally voiced sounds, because the glottal pulses produced by the alternate ends of the vibrating glottis do not occur exactly evenly interspaced between one another. So the pitch depends on the mode of vibration of the glottis as a whole, which is usually fairly slow whenever there is laryngealized voicing.

Laryngealized voicing need not occur in implosive consonants; and equally it can occur without the downward movement of the larynx which must (by definition) be present in an implosive. We can therefore separate out two kinds of glottalized consonants: what we are here calling voiced implosives (as in Igbo and Kalabari), in which there is always a downward movement of the glottis – and there may or may not be laryngealized voicing; and what we are here calling laryngealized consonants (as in Hausa), in which there is always a particular mode of vibration of the vocal cords – and there may or may not be a lowering of the larynx.

Laryngealization occurs in Hausa ʔb, ʔd and ʔj (in the orthography ɓ, ɗ, 'y) and in Margi and Bura in ʔb, ʔbd̪, ʔd, ʔw and ʔj. Most authorities agree that in each of these languages the laryngealized series of consonants differs from the series of similarly articulated voiced consonants in a uniform way. It is easy to see that in the case of the consonants ʔw and ʔj, the difference is in the mode of vibration of the vocal cords; these sounds are never literally implosive. My observations indicate that the same mode of vibration of the vocal cords occurs in the other consonants ʔb and ʔd; these sounds may be incidentally implosive on some occasions; but they are always distinguished from their voiced counterparts by being laryngealized.

The variation in the rate of glottal impulses may be seen in Plate 1b which shows the waveform which occurred in the beginning of the second syllable of Hausa ʔjáːʔjáː 'children'. The vocal cords were apparently tightly closed for at least 30 msec in between the two syllables (since there is no speech wave form for this length of time); then, when they did start vibrating, there were four glottal pulses irregularly spaced in a little under 20 msec; these pulses were followed by a gap of almost 17 msec; the next pulse was the first of a series recurring at regular intervals of about 12 msec. During some of the 17 msec before the regular vibrations began the vocal cords must have been held together; I have no criteria for deciding whether the vocal cords were together for long enough for this part of the sequence to be called a glottal stop.

It is often not possible to make an absolute distinction between laryngealization and glottal closure; as in the case of so many other phonetic oppositions there is an infinite gradation between the poles of the two categories. A familiar example is provided by British English of the kind known as RP. In this accent one of the phonetic features of some final junctures is laryngealized voicing in which the movements of the vocal cords become slower and slower so that the last one, in which the vocal cords are held together for an appreciable length of time, might well be called a glottal stop. Similarly I am in doubt concerning categorization of nearly all the 'laryngealized' consonants recorded in two of the West Atlantic languages which were investigated (Serer, and four different dialects of Fula). In these languages laryngealization typically consisted of the quickest possible change from normal voicing to a glottal stop, followed by a rapid change back to normal voicing again; the glottal stop usually occurred a few milliseconds before the consonant closure was made. In the West Atlantic languages generally the 'creaky voice' or laryngealized mode of vibration of the vocal cords is less noticeable than in the Chadic languages such as Hausa and Margi. Because I have not been able to distinguish consistently between voiced consonants with an accompanying glottal stop and similar consonants marked by laryngealization, I have symbolized both by a prefixed ?. In view of the unitary nature of the sound, in many cases the ? might well have been linked to the following sound by a tie bar.

4

STOP AND AFFRICATE ARTICULATIONS

Some of the most well-known West African stop consonants, the labial velars k͡p and g͡b, have been discussed already in Chapter 2. The present chapter reviews most of the other stop consonants which are to be found in these languages. We will consider these sounds in accordance with their place of articulation, starting in the front of the mouth and then working back.

There is little to be said about the bilabial stops p and b. All the languages investigated had some form of bilabial stop: but seven of them lacked the contrast between p and b. None of the languages investigated had a bilabial affricate.

One of the most unusual sounds recorded in the course of this investigation was the labiodental flap which occurs in ideophones in Margi. (Strictly speaking this sound is not a stop or affricate, and therefore does not belong in this chapter. But it is clearly quite different from the lingual flaps which we shall be considering later. Because it includes a stop component, it seems appropriate to describe it here.) This sound is illustrated in Plates 7a and b; the two photographs and the spectrogram are all of different utterances of the same word. The articulation consists of drawing the lower lip back inside the upper teeth, and then allowing it to flap against the teeth as it returns to its normal position. There are three stages with different acoustic correlates as numbered on the spectrogram in Plate 7b. The first stage consists of a stop with the lower lip tensed against the upper lip and teeth. The spectrogram shows that there is some voicing in this stage, which lasts for a considerable period – about 180 msec. In the next stage the stop closure is broken by pulling the lower lip back inside the upper teeth. The left-hand photograph in Plate 7a illustrates the tension of the cheek muscles involved in this stage; and the spectrogram shows that there is a vowel-like sound with little or no friction, which lasts for 80 msec. The third stage consists of the flap against the upper teeth, as indicated in the right-hand photograph; the spectrogram shows that the contact is very brief, lasting, as is the case with most flaps, for not more than 30 msec.

We must next consider the dental and alveolar stops. Here the difficulties of classification again become great. We have to consider articulatory contacts with centres which range from the tip of the upper front teeth to the back of the teeth ridge. In forming these articulations the extent of the contact area ranges from a thin strip to a wide band including both the upper teeth and much of the teeth ridge. The part of the tongue making contact with the upper surface of the mouth may be the tip, or the blade, or the whole of the tip and blade together. All these possibilities were not considered in Chapter 1, which was illustrated by a somewhat simplified chart and diagram. Similarly the transcriptions in Table 1 and Appendix B do not take this range of possibilities into account. A better tabulation with a larger array of symbols suitable for use in the next few paragraphs is given in Table 4.

The symbols used in this table are based on current I.P.A. usage; but it should be noted that (as in the 'Africa' alphabet) ɖ is used to indicate an (apical) post-alveolar stop which sounds slightly different from the retroflex stop found in Indian languages such as Hindi; this kind of retroflex stop does not occur in any of the languages considered here. Other symbols might equally well have been used for some of the other categories.

TABLE 4. Dental and alveolar articulations

(a) Part of tongue used as active articulator	tip	tip and blade	tip	blade	tip	blade
(b) Part of roof of mouth used as passive articulator	teeth	teeth and teeth ridge	front of teeth ridge		back of teeth ridge	
(c) Label summarizing (a) and (b)	(apical) dental	(laminal) denti-alveolar	apical alveolar	laminal alveolar	(apical) post-alveolar	(laminal) pre-palatal
(d) Example: phonetic symbol	d̪	d⁺	d	d-	ɖ	d̠
(e) Example: languages	Temne	Isoko Ewe	Twi Isoko	Temne	Ewe	Twi

It was only an arbitrary decision to symbolize the pre-palatal position by a symbol indicating a retracted alveolar (d̠), rather than an advanced palatal (ɟ⁺).

All these possibilities are needed if we are going to compare the pronunciation of different West African languages. It is certainly not sufficient to operate simply with the contrast dental as opposed to alveolar. Both Temne and Isoko have two contrasting stops in the region being considered; but in Temne the one made further forward in the mouth is articulated with the tip of the tongue, and the one further back involves the blade of tongue; whereas in Isoko precisely the reverse is true. Consequently the one that is called interdental in Temne (Wilson, 1961) has much in common with the apical alveolar in Isoko. The similarity of the auditory and acoustic effects of these two sounds is increased by the fact that in both languages the consonant made with the tip of the tongue (Temne t̪; Isoko t) is unaspirated; whereas the blade of the tongue consonant (Temne t-; Isoko t⁺) is aspirated and affricated in both cases.

Isoko is the only West African language that I know of that has both voiced and voiceless consonants of this kind. Temne and other Sierra Leonean languages contrast only the voiceless stops t̪ and t-; and Ewe and some of the Central Togo languages contrast only the voiced stops d⁺ and ɖ. Spectrograms of Isoko d⁺ and d are reproduced in Plate 8b. There are clear differences in the formant transitions which are due to using the blade as opposed to the tip of the tongue in the (laminal) denti-alveolar as opposed to the apical alveolar articulations. The joining of the third and fourth formants at the start of the stop release, and the unusual second formant which can be seen in the first part of the (laminal) denti-alveolar closure, occurred in a significant number of utterances of this word by the informant. The affrication which occurs with the blade articulation may also be seen.

The distinction between d⁺ (orthographic d) and ɖ in Ewe has been described by Berry (1951b). He claims that d is made with the tip of the tongue on the teeth whereas in ɖ the tip of the tongue touches the alveolar arch. He also suggests that 'The tongue is relatively flat and spread in making d . . . in pronouncing ɖ the tongue is somewhat contracted, i.e. there is a slight grooving of the blade . . . Finally . . . d is pronounced with strong force and noticeably greater tension of the muscles of the articulating organs.' I examined the contrast between d and ɖ in two dialects of Ewe (Kpandu and Peki) and in some of the neighbouring Central Togo languages, and could find little evidence to support Berry's descriptions. My own direct observations and instrumental records from the use of palatography lead me to the conclusion that in all these languages the principal difference between these two sounds is in the part of the tongue that is used. Orthographic d is articulated with the blade of the tongue against the teeth and alveolar ridge, whereas ɖ is articulated with the tip of the tongue against the alveolar ridge (usually, but not always, the posterior part). Plate 8a shows palatograms of an Ewe (Kpandu) speaker saying the words é dà 'he throws' and é ɖà 'he cooks'. It will be seen that the area of contact between the tongue and the roof of the mouth is smaller in the second phrase than in the first. Examination of the subject's tongue after the pronunciation of each phrase also made it clear that in the case of é dà the black medium had been wiped off by the tip and blade, but in é ɖà only a small part of the tip about 5 mm across had touched the roof of the mouth; since the area of contact on the roof of the mouth is wider than 5 mm the tip of the tongue must have moved as it made contact. Spectrograms for these words are shown in the plate. The acoustic difference produced by these slightly different articulations is very small, consisting mainly of a slower and more complicated transition of the second formant in é ɖà.

A completely different kind of articulation on the posterior part of the teeth ridge occurs in some Ghanaian languages. In sounds written in the local orthography as dw and tw (as in the name of the language, Twi) the stop closure is made by the blade or even the front of the tongue, the tip being down behind the lower front teeth; the centre of the tongue is raised towards the hard palate, so that the sounds may be strongly palatalized. In most of the languages in which these sounds occur there is a contrast between a labialized affricate articulated in this way, and a non-labialized affricate with a similar but not necessarily identical articulation. Thus my main Fante informant had the same mid-point of articulation in the labialized consonant in ɔd͡ʒʷè 'he calms' as in the non-labialized consonant in ɔt͡ɕè 'he catches'; but the extent of the contact was greater during the labialized consonant. Conversely, my Akwapim Twi informant had the same centre point of articulation in ócɥà 'he cuts' and càcà 'mattress'; but in this case the contact area was greater in the non-labialized consonant. In Nzima I found the articulations to be identical in ɔcɕè 'he divides' and ɔcɕʷɛ̀ 'he pulls'.

The contrast between the labialized and the non-labialized forms of these affricates usually carries a low functional load, and in some of these languages may be non-phonemic; the non-labialized form often occurs mainly before unrounded vowels. I am told (by Paul Schachter) that all these sounds are historically fronted palatals rather than affricated post-alveolars. In many of the languages which I investigated I was in doubt whether to classify these sounds as post-alveolars or pre-palatals; the distinction is, of course, largely arbitrary.

The techniques used for investigating these sounds may be illustrated by reference to the records for Late, a Guang language with obvious affinities to Twi. Plate 9 provides data on the pronunciation of the word édɥò (in the local orthography edwo) 'a yam', which contains a labialized and palatalized pre-palatal affricate. The informant in this case was very cooperative, and I managed to make an excellent impression of the roof of his mouth; this enabled me to draw both an accurate sagittal section and also a number of contour lines of the mouth sectioned in the plane of the teeth. One of these contour lines has been superimposed on the palatogram in the plate. From a consideration of the contact area in relation to this line, the shape of the centre and back of the tongue has been deduced. I also photographed the tongue after this utterance; the resulting lingua-gram, shown in the plate, shows that the tip and blade played no part in the articulation of the consonant, since the black marking medium does not appear on the anterior part of the tongue. The photograph of the lips shows that they come very close together.

The acoustic effect of all these articulations is shown in the spectrogram, which also illustrates three other points: the lack of voicing during part of the consonant; the small amount of affrication which occurs after the release of the closure; and the comparatively lengthy offglide involving a semi-vowel of the ɥ type.

It is interesting that a number of neighbouring Ghanaian languages, such as Gã, Effutu, Late and Anum, do not make a contrast between dˈ and ɖ, but nevertheless have two different articulations: a blade denti-alveolar form for the voiceless stop tˈ, and a tip alveolar or post-alveolar form for the voiced stop d or ɖ as in Ewe.

There is often a great deal of variation in the application of the term palatal. Some authors have used it to specify what we have here called pre-palatal (as in the Ghanaian languages discussed above); and others use it to include what we would call palatalized velars (such as the kj for c in some forms of Hausa kjàŋʷá 'cat', where the contact is in the velar region but the front of the tongue is also raised towards the hard palate). As the material in this monograph continually forces us to remember, there are no absolute categories of place of articulation.

Comparatively few West African languages form stop consonants in the middle of the area we have called palatal (see Figure 2); and most of these languages have palatal affricates rather than plain stops. No West African language that I know of contrasts these two possibilities. The only language investigated which contrasts palatal stops and the closely similar post-alveolar affricates is Ngwo, which has a stop system including d, dz, dʒ, ɟ, and g. The middle three terms in this series are illustrated in Plate 10. This was another occasion when I was able to make accurate observations of the informant's pronunciation by means of palatography combined with information about the vocal organs obtained from a dental impression of the roof of the mouth. The pictures show that there are three distinct places of articulation. (Other pictures show that g is also distinctly different; but the contact areas for d are the same as those for dz.) It should be remembered that palatograms are apt to be misleading in that they often show only the contact between the edges of the tongue and the roof of the mouth. They do not indicate the possibility of a groove or a hollow in the middle of the tongue which may be important in the formation of fricative sounds (Ladefoged, 1957).

The velar stops in West African languages require little comment here. All the sixty-one languages had a voiceless velar stop; and most of them had a contrasting voiced counter-part. Two or three languages, such as Limba, had k in morpheme initial position and either g or ɣ in other positions. No language used a velar affricate in contrast with a velar stop.

Serer uses a contrast between a uvular and a velar stop, illustrated in Figure 6 by trac-ings of the formant patterns in four Serer words. Transitions during the release of these

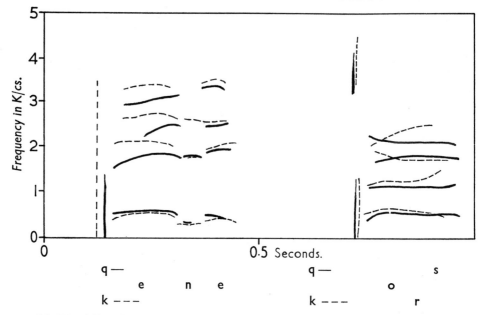

M. Diouf (Serer): qene, kene, qos, kor 'This boy, that, leg, man' 9 April 60

FIG. 6. Superimposed tracings from spectrograms illustrating the difference between uvular and velar stops in Serer.

aspirated stops were not very noticeable. The major difference between q and k in these words is in the lower position of the second formant of the vowel after the uvular stop.

Some Chadic and West Atlantic languages use a glottal stop as a regular phoneme in contrast with other stops. In addition, several different West African languages (e.g. Tiv, Efik, Igbira) are strikingly similar to English in one small feature of their phonologies: the glottal stop is used contrastively only in the word (interjection?) meaning 'no' which always has the form áʔà or ʔáʔà. It would be interesting to know the extent to which this correspondence is paralleled throughout the languages of the rest of the world.

5

NASALS AND NASALIZATION

Some West African languages have a larger number of nasal phonemes than are commonly to be found in European languages. Idoma has six contrasting items m–n–ɲ–ŋ–ŋʷ–ŋ͡m all of which pattern as unit phonemes in the language. The labial velar ŋ͡m, which was discussed and illustrated on page 12, is clearly a single unit. The labialized velar in the series might be regarded as a sequence; but this would lead to a slightly eccentric phonotactic statement, since ŋʷ would be almost the only permitted consonant cluster. Even on this alternative analysis Idoma still has five nasal phonemes differentiated solely by position of articulation. This is one more than the maximum reported by Hockett (1955). A few other languages also have systems with five nasal consonants; thus both Dagbani and Gã have the labial ŋ͡m as well as m, n, ɲ, ŋ. Conversely, since nasalization often operates as a prosodic feature of a whole syllable, in many languages it is possible to recognize a comparatively small number of nasal consonants, provided the prosody of nasalization is attached to the vowel and the contrast between oral and nasal vowels is marked. In accord with the view outlined by Hockett (1955), some languages may have only one nasal consonant in a phonemic transcription.

In Yoruba, when the consonants w, r, j occur before nasal vowels, the nasalization of the syllable as a whole is readily apparent; the words ɛ̀wɔ̀, ɛrɔ̄, ójɔ́ 'chain, meat, he yawned' are, in a narrow allophonic transcription, ɛ̀w̃ɔ̀, ɛɹ̃ɔ̄, ój̃ɔ́. Some writers (e.g. Ward, 1952) were so struck by the distinctive qualities of the sounds w̃ and j̃ in Yoruba that they set them up as separate phonemes. But as Siertsema (1958b) has pointed out, there is no justification for doing this, as the nasalization is completely predictable.

In Bini and other Edo languages j and w in nasal syllables are allophonically ɲ and ŋʷ respectively, both having complete contact between the tongue and the roof of the mouth as shown by palatography. Interestingly enough the local orthographies consistently mark nasality by adding an n after the vowel; but they nevertheless recognise that the nasality feature extends over the syllable as a whole, including the consonant before the vowel where appropriate. In contrast with linguists such as Melzian (1937), native speakers of Edo languages do not find any need for separate letters to represent the palatal and labialized velar nasals. In Ora, an Edo language of the Northern part of Benin Province, the word for 'every evening', which can be represented in a phonemic transcription as éwẽwà̰ and in a narrower allophonic transcription as éŋʷẽ̃ŋʷà̰, is written locally as ewenwan; and the word for 'albino', which is phonemically ájárí, allophonically áɲã́ɹ̃í, is written as ayanrin.

Among the pairs of oral and nasal sounds which are distributed in this way in some languages are: ɹ–ɹ̃; j–j̃ or ɲ ; w–w̃ or ŋʷ; n–l. In order to show that there is a distinction in certain languages between oral and nasal vowels, and between the members of one of these pairs, it is necessary to find contrasts between at least three out of the four phonetic items CV–CṼ–NV–NṼ (where C stands for the oral consonant in the suspicious pair (Pike, 1947), N for the corresponding nasal or nasalized consonant, and V for one of the vowels of the language). As far as I can see, on this basis there may be no reason for regarding l and n as distinct phonemes in Yoruba and other related languages. There are stems which appear in the written language as -na and -la; but the written language does not have *lan or *nan. The vowel in na is nasalized; so this might well be phonemically lã, just as w̃ũ or ŋʷũ is phonemically wũ. On this analysis Yoruba has only the one nasal phoneme m. This runs counter to Ferguson's tentative assumption (in Greenberg, ed. 1963) that 'If in a language there is only one P(rimary) N(asal) C(onsonant) it is n.' Yoruba is probably not the only West African language that contradicts this hypothesis.

Several widely different languages (Serer, Fula, Mende, Sherbro, Tiv, Kutep, Margi) have nasal compound stops contrasting with both voiced stops and nasals. In these languages these sounds are usually unit phonemes, not sequences of syllabic (tone-carrying) nasals followed by syllable initial stops. In Tiv the initial nasal component clearly patterns like an additive component such as voicing, which can occur at each place of articulation in the stop consonant series. The stop system of Tiv includes the six sets: p, b, m; k͡p, g͡b, ŋ͡m; t, d, n; ts, dz, ndz; tɕ, dʑ, ndʑ; k, g, ŋg. But there are only the three nasals: m, n, ɲ.

Both Ferguson (in Greenberg, ed. 1963) and Hockett (1955) also suggest that no language has nasals articulated at more places than stops. Igbira has four places in the stop series p, t, c, k but five nasals m, n, ɲ ,ŋ, ŋʷ. It is true that the functional load of the opposition ŋ–ŋʷ is very low. The sound ŋʷ is quite common, but I know of only three morphemes containing ŋ. Nevertheless these three items do not seem to be explicable in any way except by setting up ŋ as a separate phoneme. Among the stops it is the non-labialized forms k and g which occur regularly; I am fairly certain that kʷ and gʷ never occur in the lexicon (although they do occur in combinations of morphemes containing an elided u).

6

FRICATIVES AND APPROXIMANTS

The term *approximant* is used here to describe a sound which belongs to the phonetic class vocoid or central resonant oral (Pike, 1943), and simultaneously to the phonological class consonant in that it occurs in the same phonotactic patterns as stops, fricatives and nasals. The bilabial and labiodental approximants and fricatives which occur in West African languages include a number of features not common in other languages. There is, for example, a contrast between bilabial and labiodental fricatives in Ghanaian languages such as Ewe. Photographs illustrating the difference in one of the neighbouring languages, Logba, are reproduced in Plate 16*b*. Both lips are tensed and almost touching for the bilabial fricative; in the labiodental not only is the lower lip drawn back so that (in most cases) it actually touches the upper front teeth, but also the upper lip is actively raised up from its rest position. (This raising of the upper lip is even more noticeable in side view photographs of other informants.) All the informants of Ewe, Logba and Avatime (Siya) pronounced these sounds in essentially the same way. Spectrograms of two Ewe sentences designed to illustrate the differences between voiced and voiceless, bilabial and labiodental fricatives in near minimal pairs are shown in Plate 11. The informant was conscious of the reason for recording these sentences, and consequently the fricatives may be somewhat longer than usual. Nevertheless the difference between the two sounds in each pair is very small. In each case the point of origin of the second formant is slightly lower for the vowel following the bilabial fricative.

Ewe and Avatime (but not Logba) have a voiced labial velar approximant w in addition to the bilabial and labiodental fricatives β and v; but none of these languages has voiceless counterparts to all three sounds. Ewe and Logba have a pair of voiceless fricatives ɸ and f, but no voiceless labial velar ʍ. Avatime has f and a voiceless labial velar ʍ. Some of the neighbouring Akan languages (Krachi, Anum, Late and Nkonya; but not Twi and Fante (see page 28) have a contrast between an approximant ʍ and f and w; but again they do not have the three possibilities in the voiceless series, since they all lack ɸ. One or two languages (Nupe) have ɸ but no f. Serer, Dyola and the Fula of Mopti have neither v nor w, but a bilabial or labiodental approximant here symbolized by ʋ. This sound differs from v mainly by the lack of friction, and from w by the absence of the raising of the back of the tongue.

In some languages there is a voiced bilabial approximant ʊ in contrast with a labial velar fricative or approximant or both. The I.P.A. does not provide for the symbolization of the contrast between a labial velar approximant like the English w, and a similar sound with a closer articulation which may produce audible friction. This latter sound is here symbolized by w˕. At least four Edo languages have the three sounds w˕, ʊ, v. Plate 12 shows photographs of the lip positions in the consonants in the three Isoko words ɛʊɛ 'breath', ɛvɛ 'how' and ɛw˕ɛ 'hoe', together with a photograph taken during the vowel in the first word. The two views, full-face and side-face, were taken simultaneously with the aid of a mirror. As explained in the Introduction, the moment in time when each photograph was taken was checked by noting the time of occurrence of the click of the camera shutter on the recordings, which were also made simultaneously. This shutter noise mars the recordings of these words from the point of view of spectrographic analysis. Accordingly another recording was used to make a set of spectrograms, which are reproduced in Plate 13a. The data in these two plates enable us to describe these contrasts in some detail. In the spectrograms of wᴸ the large changes in the frequencies of the formants show that the consonant and the vowels must have different lip and tongue positions; neither a lip nor a tongue movement alone could have produced a lowering of the second formant of this magnitude, even though, as the photograph of wᴸ shows, the protruded lips came very close together. The drop in intensity in the spectrograms during this sound also indicates a close articulation. The small changes in the formant frequencies in ʋ indicate that there is only a very small difference in the articulation throughout this word. Comparison of the first and second photographs in Plate 12 shows that there is a difference in the lip positions of ɛ and ʋ, but, as far as can be seen, the tongue is in much the same position in the consonant and in the vowel. The constancy of the intensity on the spectrogram can also be correlated with the fact that the lips did not come very close together. In the word évé there is a considerable drop in the intensity during the consonant. This is in accord with the photographic data which show the lower lip pressed against the upper teeth during the articulation of the v. The formant transitions are much the same in v and ʋ so we may conclude that in v, as in ʋ, the tongue was probably in the same position for the vowel as for the consonant.

Bini ʋ has a slightly closer approximation of the articulators than occurs in the corresponding sound in Isoko. Even so, Bini ʋ is usually an approximant rather than a fricative. Incidental evidence on this point is provided by Plate 3a. The oral pressure recording shows that the lips did not come very close together, since there is only a small increase in the mouth pressure during the articulation.

The similarity (to European ears) of the sounds v, ʋ, wᴸ is further complicated by the fact that both Urhobo and Isoko have a second voiced labial velar more like the English w, in addition to wᴸ. These two languages also have an unrounded velar fricative ɣ. The corresponding articulation in Bini is more of an approximant than a fricative, and may be symbolized ɣᵀ. My somewhat inadequate data indicate that there is probably no velar fricative or approximant in Ora. In summary, therefore, the distribution of these phonemes in the four Edo languages we have been considering is as shown in Table 5. Words illustrating these contrasts are given in Appendix B.

TABLE 5. Some contrasting items in Edo languages

	fricative		approximant	
	symbol	languages	symbol	languages
bilabial	β	—	ʋ	B, I, O, U
labiodental	v	B, O, I, U		
velar	ɣ	I, U	ɣᵀ	B
labial velar	wᴸ	I, U	w	B, O, I, U

B=Bini, O=Ora, I=Isoko, U=Urhobo

Two other languages, Bura and Margi, also have a contrast between fricative and approximant labial velars wᴸ and w. An example of this phonemic contrast in Bura is illustrated in Plate 14. As would be expected, the closer approximation of the articulators causes a more noticeable decrease in intensity and a far larger second formant movement in the case of the fricative variety. The high-frequency random energy usually associated with a fricative noise source is not evident. This is probably because the velar closure is the greater of the two, and the slight amount of local turbulence in the airflow which occurs in a voiced velar fricative with a low rate of flow is insufficient to be recorded when there is a close approximation of the lips. Both Bura and Margi also have a voiceless labial approximant ʍ; but they do not have a distinction between a fricative and an approximant of this nature. They fall in with the general rule, previously exemplified in the case of the Ghanaian languages discussed earlier in this chapter, that there are apt to be fewer distinctions among voiceless than among voiced sounds of this sort. This can presumably be related to the smaller auditory difference between the voiceless sounds.

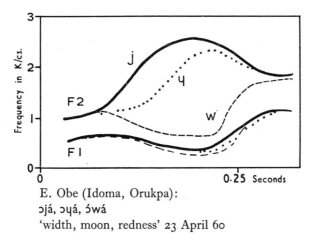

E. Obe (Idoma, Orukpa):
ɔjá, ɔɥá, ʃwá
'width, moon, redness' 23 April 60

FIG. 7. Superimposed tracings of the first and second formants in Idoma palatal, labial palatal, and labial velar approximants.

A number of languages have a contrast between the labial velar and the labial palatal approximants w and ɥ. Among the Ghanaian languages such as Twi, Fante, Nzima and Gã, the contrasts may not be phonemic if appropriate grammatical considerations are taken into account. But in Western Idoma there is a clear contrast, illustrated in Figure 7, which shows superimposed tracings of the first and second formants in the words ɔjá, ɔɥá, ʃwá, 'width, moon, redness'. The differences in the first formants are small. The position of the second formant for ɥ is clearly intermediate between that for j and w, indicating that the rise due to the high front movement of the tongue is simultaneously lessened by the fall due to the rounding of the lips. It is also apparent that in the w the rounding of the lips and the upward movement of the tongue take longer than the release of these articulations.

Dental fricatives are uncommon in West African languages. Sumner (1922) states that a voiceless dental fricative occurs in Temne; but, as we have noted earlier, the contrast is really between two kinds of stop, rather than between a stop and a fricative. A nearby Sierra Leone language, Sherbro, does have θ contrasting with tˢ. This is the only example of a dental fricative in the sixty-one languages considered in this study.

Throughout this book the symbols ʃ and ɕ have been used to distinguish between two organically different fricatives, the one being made with the tip of the tongue near the post alveolar ridge, and the other involving the blade of the tongue and the prepalatal area. None of these languages distinguishes between these two sounds. But some of my Twi informants consistently used these two different articulations together with added lip rounding to form a phonemic contrast. The word ɔɕé 'boundary' is usually pronounced with the tip of the tongue down behind the lower front teeth; but the word ɔʃʷɛ̀ 'he looked at' often has the tip of the tongue raised, sometimes almost into the retroflex position. Both Twi and Fante have this rounded fricative in words where other Akan languages have a voiceless labial velar ʍ.

The voiced palatal approximant j was found in all the sixty-one languages investigated, with the exceptions of Wolof and Gɛ̃ (and in the latter case this may simply reflect lack of data on my part). Hoffmann (1957, 1963) states that Bura and Margi have a contrast between a palatal fricative and a palatal approximant. The *Principles* (1949) of the I.P.A. say that there is no separate symbol for the fricative and frictionless varieties of j because they have not been found as separate phonemes in any language. I did not note this contrast in Bura (considering the usual reliability of Hoffmann's records, this no doubt simply shows my lack of observation; I did not have access to Hoffmann's analysis of Bura at that time). But I did, on the basis of Hoffmann's word lists, record the contrast in Margi. In general the difference is similar to the contrast between the fricative and approximant wʌ and w in these two languages. There is a larger movement of the second formant and a greater drop in intensity in jʌ, due to the closer approximation of the articulators. The fricative component is more clearly audible in jʌ than in wʌ, because of the spread lip position.

Only Bura, Margi, Igbo, and Late were recorded as having a voiceless palatial fricative ç. In the case of Late this sound occupies the place in the consonant system where other similar languages have ʃ or ɕ.

Velar fricatives were only slightly more common. Bini and Gɛ̃ were the only two languages to have both voiced and voiceless ɣ and x as well as h (in free variation with ɦ). Nzima, Isoko and Urhobo have ɣ in contrast with h which (as often) is in free variation with x; and Ngwo and Bafut (apparently, my data are not very good) have ɣ but no x or h. A few languages (Efik, Ijaw) have ɣ simply as an intervocalic allophone of k or g.

Voiceless uvular fricatives were found in Soso, Serer and Wolof. The Wolof χ was illustrated in Plate 6. The voiced uvular fricative ʁ occurs in Krio wherever the corresponding English words have r. Dagbani has ʁ as an allophone of g after short vowels.

Clusters involving w, f, v, ɣ, and other sounds as additive components will be discussed later (Chapter 8).

7

LATERALS AND FORMS OF r

The most complicated series of laterals in the sixty-one languages investigated were those in Bura and Margi. Both these languages have a voiced alveolar lateral approximant l contrasting with a fricative ɮ, as well as a voiceless alveolar lateral fricative ɬ. In addition to these three sounds Bura has a voiced palatal lateral approximant ʎ and a voiceless palatal lateral fricative ʎ̥, making a total of five lateral phonemes. The only other language in the sample with more than one lateral is Kom, which has ɬ and l. The Bura alveolar lateral fricative was illustrated in Plate 14. The fricative noise has a fairly low intensity, and seems to be composed of two bands of energy centred at about 2,400 and 4,400 cps.

Twenty of the languages investigated have no phonemic contrast between l and some form of r. Some languages have only one articulation of this kind (usually l). Many others have ɹ and l in free variation. But some languages, such as Ewe, have both a lateral (approximant) and a central articulation which may be an approximant ɹ or a tap ɾ, the two possibilities (lateral and central) being in complementary distribution. In Ewe l occurs intervocalically and after all consonants except alveolars and palatals. Some Ewe alveolars (ɖ, n, z, l) can be followed only by a vowel; but the other alveolars (t, ts, d, dz, s) and the palatals (ɲ, j) can be followed by a vowel or ɹ/ɾ. In many of the languages that do have both l and some form of r, there is often a great deal of free variation in the actual articulation of the latter sound. Thus Temne has apparent free variation between r, ɹ and ɾ.

Some of the Edo languages have a four-way contrast between the three voiced sounds l, ɹ, and ɾ, and an additional voiceless sound ɹ̥. Plate 13b shows three Isoko words contrasting the three voiced sounds. The similarity between l and the Isoko kind of ɹ is very great. The onset transitions are almost identical. The major difference is in the intensity of the third formant during the stricture, and in the abrupt increase in energy in the higher frequencies after the articulation of l. There are also some small differences in the offglide transitions of the third formant. All these factors would indicate a similar movement of the tip of the tongue, perhaps combined with a hollowing of the centre or raising of the back of the tongue for both sounds, and with certainly less articulatory contact for ɹ than for l. The one tap alveolar ɾ differs from both these sounds in two ways. Firstly there is a difference in the timing of the movement. Secondly there is an important difference in the formant pattern, in that the third formant remains fairly constant during the formation of the articulation. We may conclude that this tap is formed by an upward movement of the tip of the tongue, similar (except for its timing) to the gesture involved in the formation of d, and without the hollowing or raising of the back of the tongue which occurs in Isoko r and l. This conclusion is supported by palatographic evidence.

The description of the sounds of Bini by Melzian (1937) is more or less in accord with the description of the Isoko contrasts which have just been given. Melzian refers to l, trilled r (corresponding to the tap ɾ above), and 'a sound intermediate between r and l', which is certainly a good description of this particular approximant. My Bini and Ora informants, however, did not use the difference between a tap or trilled r and an approximant ɹ with an l-like quality; instead they distinguished between a fricative ɹ̝ and an approximant ɹ. The I.P.A. *Principles* (1949) do not consider the possibility of a phonemic distinction of this kind. Very often the Bini and Ora speakers augmented the distinction between these two sounds by making the approximant with the smallest possible articulation, so that the effect is almost of the absence of any consonant between the first two vowels in words such as áɹába 'rubber'.

The fricative ɹ˔ might be identified with the sound symbolized by ʒ in descriptions of other languages. Bini has no ʒ; and in our system of classifying consonants in terms of a small number of dimensions, the two symbols ɹ˔ and ʒ occur at the intersection of the same set of categories. They both represent voiced apical post-alveolar fricatives. The difference is in the position of the part of the tongue immediately behind the tip: for ʒ this is either flat or actually raised towards the hard palate; for ɹ˔ it is hollowed.

Alveolar tap ɾ also occurs in Hausa. Most of my informants used this sound, although occasionally it was replaced by a trilled r. Both these sounds occurred where Hausa orthography has r. There is also a contrasting sound which is sometimes written as r, and which appears to be a retroflex flap ɽ. Most authorities (James and Bargery, 1925; Greenberg, 1941; Hodge, 1947) state that the difference between these two sounds is that the one is a trill and the other is a flap. The nearest I can come to agreeing with this is to say that the first sound is a trill which has a statistical probability of consisting of only one tap. The typical forms recorded in the present investigation are illustrated in the spectrogram of the Hausa utterance reproduced in Plate 15a. The acoustic similarity is obvious. Indeed I have not been able to find any consistent acoustic difference between the two sounds. The second formant is slightly higher in the first word than in the last in Plate 15a; but this does not occur in other spectrograms of the same words. There must, of course, be some acoustic cues, probably in the formant transitions. If there were not, Hausa speakers could not learn to be consistent in making the different articulations which are clearly revealed by palatography. Plate 15b shows palatograms of the two words baɾa and baɽa, each spoken in isolation. Above each photograph is a cross-section of part of the vocal tract, based on tracings of an impression of the roof of the informant's mouth. This section has been lined up so that the frontal incisors are above the corresponding teeth in the photograph. So we may be sure that a contact area which appears to be a certain distance behind the front teeth on the palatogram, indicates a point of contact as shown in the diagram above. The curve of the tongue is deduced from observations of the contact areas on the sides of the palate, and from knowledge of the contours of the palate obtained from the impression. We cannot, of course, be sure of the direction of the movement. But the width of the contact area in each case suggests that it was as indicated. (Movements of this sort are the only way that I, imitating this informant, can produce narrow contact areas of this kind.)

These two articulations are very different from the speaker's point of view. The ɾ involves a rapid movement of the tip of the tongue up to tap (and occasionally to trill) against the forward part of the alveolar ridge. The ɽ is made by drawing the tongue tip up and back, and allowing it to flap against the posterior part of the alveolar ridge as it comes down. Despite their acoustic similarity, which is no doubt partly due to the rapidity of the movements, it is not surprising that they constitute different phonemes.

8

S,ECONDARY ARTICULATIONS AND CLUSTERS

So far, apart from a few remarks such as those on labialization in the Akan languages (page 20), and on r and l in Ewe clusters (page 29), we have been considering mainly unit phonemes with no secondary articulations. By far the majority of languages considered here have no consonant clusters, and can be described in terms of very simply phonotactic structures. In many Kwa languages, for example, most verb stems begin with a consonant and have the form CV or CVCV, and most nouns begin with a vowel and have the form VCV or VCVCV. There are, however, a number of languages in which it might be convenient to specify at least secondary articulations or additive components such as palatalization. Carnochan (1948) has shown that Igbo verb stems can be treated in this way. Palatalization also occurs extensively in languages of different groups such as Birom and Tiv; and Birom and Ngwo have series of labialized consonants. Velarization is less usual. In Kyerepong s, n, l were recorded as being velarized, but these were the only examples found in the course of this study.

Palatalization, labialization and velarization are the most well known secondary articulations. (Laryngealization and nasalization, which are not regarded as articulations, have been treated elsewhere – pages 16 and 23). But these are not the only possibilities. Some stop systems also have addèd fricative components which are not homorganic with the stop articulation. In these cases it is difficult to know whether to regard the sounds as consisting of a sequence of consonants, or as a combination involving an additive component. Kom has the velarized forms bɣ, dɣ, which are clearly sequences from the auditory point of view; but equally the articulatory gestures overlap, in that the velar stricture is formed during the stop closure. In this particular language there are strong grounds for saying that this is a kind of additive component or secondary articulation, since Kom also has a labiodental fricative which seems to be superimposable on other articulations. The sounds observed in this language include tɕᶠ, kᶠ, gᵛ, jᵛ. Plate 16a shows spectrograms of two Kom words, the first of which demonstrates the coarticulation of tɕ and f, and the second the 'labiodentalization' of the whole second syllable of a word. In the first word the fricative noise occurs with a higher intensity over a wider frequency range than that caused by a single local source of turbulence in the air flow. In the second word the fricative component starts during the first consonant of the last syllable, and is evident right through the succeeding vowel and the final consonant. This is clearly a case in which a segmental phonemic transcription is apt to be misleading unless it is accompanied by appropriate rules for its interpretation.

A similar secondary articulation also occurs in Kutep; but in this language labiodentalization occurs only after fricatives (including those in affricates) and is in complementary distribution with labialization, which occurs after stops and nasals. There is a slight difficulty in regarding labiodentalization as an additive component in Kutep in that it often involves a concomitant difference in tongue position. Palatography shows clearly that in tʃ, ʃ, ʒ the tip of the tongue is raised; but the similar sounds tɕᶠ, ɕᶠ, ʐᵛ are made with the tip of the tongue down. Further evidence on this point is afforded by Figure 8, which is based on tracings from two laminagrams (X-ray pictures showing specific planes

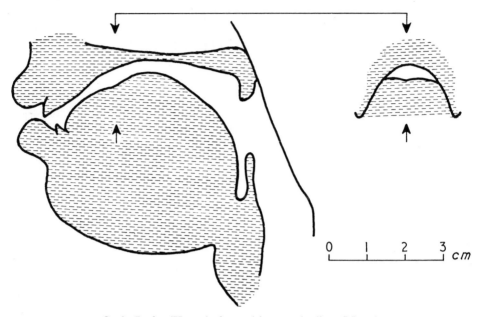

C. A. Iyaba (Kutep): baʒᵛe 'they washed' 21 May 62

FIG. 8. Tracings of X-ray photographs (laminagrams) taking during the pronunciation of Kutep ʒᵛ.

at given distances from the X-ray source, as opposed to the sum of the obstructions between the X-ray source and the photographic plate). In making these laminagrams the tongue position for ʒᵛ as in baʒᵛe 'they washed' was held for about one second. This may mean that the articulation is not entirely as in normal speech; but the position shown seems to be in accord with the palatographic data. In addition to the two primary articulations, one in the neighbourhood of the lips and the other near the alveolar ridge, there is also a secondary stricture formed by the raising of the centre of the tongue. The transverse section shows that there is a small hollow in the raised part of the tongue opposite the middle of the hard palate.

Kutep also has some other interesting combinations of consonants. There is a velar additive component which is usually a stop (and so clearly a sequence) in items such as fk, sk, skʷ; and usually a palatal or velar fricative in combinations with the stop t. But there is a great deal of free variation. I have heard an informant repeat within minutes batkà – batçà – batxà 'they burn'.

The most complicated series of consonants investigated in the course of this study occurred in Margi. On the basis of Hoffmann (1963) eighty-nine different ways of beginning a morpheme were recorded in this language. Examples of words illustrating these consonantal clusters (and the two other consonants) are given in Appendix B. It may be seen that the stop system can be characterized in terms of a number of phonetic features such as voicing, laryngealization, labialization and nasalization, some of which may occur simultaneously. In this language there is certainly a very great variety of phonetic events which have to be noted by the investigator.

9

VOWELS

Variations in vowels may be considered in terms of length, nasality and articulatory quality. Discussion of vowel length is always complicated by the interaction of the phonological analysis of length and tone. Yoruba, for example, may be considered (Olmsted, 1953) to have nine tones, three being relatively constant pitches occurring on comparatively short vowels, whereas the other six always involve a change of pitch and occur on longer vowels. The more usual description of Yoruba (Ward, 1952; Siertsema, 1959) specifies only three tones, and regards the glides as sequences of these three tones occurring on consecutive short vowels. In general it would seem that when, as in many Kwa languages, perceptually long vowels can be on one pitch or involve a change of pitch, and when these vowels can occur in the same phonological structures as sequences of different vowels, then it is preferable to regard them as being sequences of two vowels. But when, as in Fula and Hausa (with certain minor qualifications), perceptually long vowels always have a relatively constant pitch, and occur in different circumstances to sequences of vowels which may or may not have the same tone, then we may conveniently speak of the language as having long and short vowels.

Some aspects of vowel nasality have been discussed already (page 23). As in all other known languages (Ferguson in Greenberg, Ed., 1962) there were never more, and (in these languages) always fewer nasal vowels than oral vowels.

The articulatory qualities of the vowels in most of the languages were comparatively simple. They were nearly all describable in terms of a two-dimensional vowel chart in which the degree of lip-rounding is determined by the so-called tongue position (Ladefoged, 1960): front vowels were usually unrounded, as i and e; and back vowels nearly always had progressively more lip-rounding as the tongue was raised from the low central or back position of a, through o, to the high back position of u.

High back or central unrounded vowels ɯ or ɨ occur in Sierra Leonean languages such as Temne and Sherbro; and Kambari (Hoffmann, personal communication) has a distinction between two mid-central unrounded vowels which I heard as ə and ə˕ in mə́lljóó – mə́˕lljóó '*weeds' – '*weeds'. I had no opportunity of verifying the articulatory description of any of these central vowels by instrumental means.

More complicated vowel systems occur in some Eastern Nigerian and Cameroun languages such as Kom (Bruens, 1947), Bafut, Ngwo and Ngwe (Dunstan, 1964). A

TABLE 6. Some of the vowel contrasts in Ngwe

1. mbi	white chalk	8. mbu	corners
2. mbe	knife	9. aty	stick
3. mbɛ	sheath	10. mbɯ	dog
4. mbæ	pepper	11. mbɤ	brass gong
5. mba	person	12. ntsʌ	water
6. mbɔ	god	13. mbɞ	tadpoles
7. mbo	hands		

cine-radiology film was made of a Ngwe speaker saying the words in Table 6. (In addition to these words there are a number of more diphthongal vowels.)

Superimposed diagrams of the vowels in the words in Table 6 are shown in Figures 9 and 10. The positions indicated are derived from tracings of the frames in the film when the articulators approached most nearly to a steady state.

The first four vowels i, e, ɛ, æ, show approximately equal tongue movements; but it is interesting that there is only a very small difference in the lip opening and jaw positions between vowels 2 and 3. The next four vowels are here symbolized ɑ, ɔ, o, u. The reader is reminded that, although the symbols of the nearest cardinal vowels are used, the back vowels of this language should not be considered as being absolutely identical with cardinal vowels 5, 6, 7 and 8. They are, however, somewhat similar; and they appear to be roughly auditorily equidistant. It may be seen that the specification of these back vowels in terms of the highest point of the tongue does not lead to a description which is in accord with the auditory impression. The high points of the tongue for the vowels in words 6 and 7 are too close to each other; and the high points for all four vowels do not fall on any simple form of line. From the point of view of compatibility with the auditory attributes, a better form of description is that suggested by Stevens and House (1955), in which the relevant parameters are the cross-sectional area of the vocal tract at the point of maximum constriction and the distance of this point from the glottis. The cross-sectional

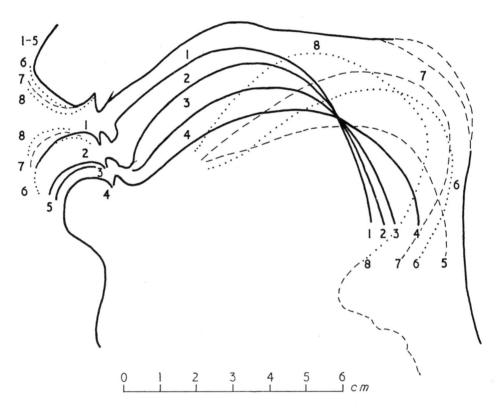

FIG. 9. Tracings from single frames in a cine-radiology film showing the tongue positions in the vowels in the first eight Ngwe words in Table 6.

34

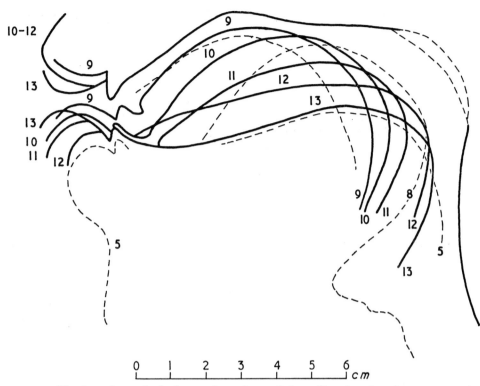

FIG. 10. Tracings from single frames in a cine-radiology film showing the tongue positions in the vowels in the last five Ngwe words in Table 6. The tongue positions of the vowels in words 1, 5 and 8 are shown by dashed lines.

area at the point of maximum constriction is very similar for all four of these back vowels; and the distance of this point from the glottis becomes larger by progressively larger steps. These two parameters plotted on logarithmic coordinates provide a useful characterization of the Ngwe back vowels.

Figure 9 might seem to indicate that the vowel in word 8 was a central rather than a back vowel. But, as Figure 10 shows, it is clearly a back vowel in comparison with the vowels in words 9 and 10, here symbolized by y and ɯ, and regarded as centralized front rounded and centralized back unrounded high vowels respectively. In addition to ɯ Ngwe has two centralized back unrounded vowels, exemplified by the vowels in words 11 and 12, and here symbolized by ɤ and ʌ. The vowel y in word 9 is the only example of a front rounded vowel. But the vowel in word 13 sounded not unlike a more open front rounded vowel such as ø or œ. Before the X-ray was taken it was very difficult to judge from the auditory cues exactly how this sound was made. In fact the tongue position is not unlike that in word 5, which was symbolized by ɑ (except that in the vowel in word 13 the point of maximum constriction in the vocal tract is slightly further away from the glottis). But despite the, basically, low back tongue position, the jaw was raised (as is shown by the position of the lower front teeth), and the lips were very close together and protruded. I do not know of the specification of a vowel position such as this in any other language.

35

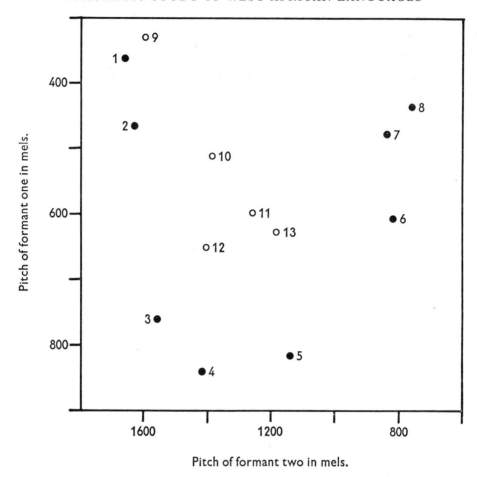

FIG. 11. Formant chart of the Ngwe vowels in Table 6.

The first two formants of these thirteen vowels are shown in Figure 11. The values indicated are derived from an estimation of the formant frequencies at the moment in the acoustic record corresponding to the frame of the cine-radiology film which was used as a basis for the diagrammed positions shown in Figures 9 and 10. Figure 11 shows the inadequacies of a simple two-formant representation of this kind. The grouping of the points does not reflect the auditory impression given by these vowels, even in the comparatively simple case of the vowels in the first eight words, all of which have a combination of lip and tongue positions which can be predicted by reference to the primary cardinal vowel plane; and the points for the remaining five vowels certainly do not convey enough information to enable the reader to reproduce a reasonable approximation of these sounds.

Some features of vowel quality cannot be adequately specified within our present terminology. The vowels of many West African languages are affected by a phonetic feature which we may for the moment equate with the tense/lax opposition. This feature usually operates at the level of the phonological word rather than the phoneme. As a result the vowel qualities which can occur within a single phonological word are severely restricted, giving rise to what is known as 'vowel harmony'. Thus in some forms of Twi (Berry, 1957) the complete range of vowel qualities is as shown in the centre of Table 7(a); these vowels can be divided into two groups as shown on the left and right of the table. In general, a Twi word of more than one syllable contains vowels of only one of these groups; so it is possible to say that the property of 'tenseness' or 'laxness' is a prosody of the word as a whole, since lax and tense vowels do not occur in the same word. The similar vowel harmony situation in one of the other Akan languages was succinctly stated by Welmers (1946), who presented an analysis of Fante in which there are five vowel qualities and a suprasegmental phoneme 'heightening' the vowels.

Nzima, Kyerepong and some forms of Twi were the only languages investigated that had two complete sets of five vowel qualities. Fante, Igbira and some other Kwa languages have a more incomplete form of vowel harmony as shown in Table 7 (b). These languages have nine vowel qualities which can be divided into two sets. The most open vowel is common to both sets, so that the opposition tense/lax does not apply at this level. An

TABLE 7. Some patterns of vowel qualities that occur in vowel harmony sets in West African languages

```
            Set 1                    Combined sets                   Set 2

           i      u                 i            u
                                      ɪ       ʊ                    ɪ      ʊ
(a)        e   o                   e        o
                                       ɛ   ɔ                       ɛ  ɔ
              a                       a ɑ                            ɑ


           i      u                 i            u
                                      ɪ       ʊ                    ɪ      ʊ
(b)        e   o                   e        o
                                       ɛ   ɔ                       ɛ  ɔ
              a                        a                            a


           i      u                 i            u
            e   o                  e        o                   i        u
(c)
              a                        ɛ   ɔ                       ɛ  ɔ
                                       a                           a


           i      u                 i          u
                                      e      ʊ                   e      ʊ
(d)        ɛ   o                    ɛ      o
                                       a  ɔ                       a  ɔ
```

even more incomplete form of vowel harmony occurs in Yoruba and Idoma, where there are seven vowels which can be divided into two sets of five as shown in Table 7 (c). When there is this amount of overlapping between the sets it seems doubtful whether it is useful to ascribe a feature of tenseness or laxness to the word as a whole.

Most of the languages displaying any form of vowel harmony have five vowels in each sub-set. But in Igbo there are two distinct sets of four vowels as shown in Table 7 (d). Like many languages which display vowel harmony in polysyllabic lexical items, Igbo also has two forms for affixes such as the pronominal verbal forms. The Igbo prefix ó ('he' past tense) is used with verbs such as sìrì, sèrè 'cook, quarrel', whereas with verbs such as sèrè, sàrà, 'say, wash', the form ɔ is used. Carnochan (1960) has shown how the description of vowel harmony can be extended still further for this language. Many other languages such as Serer (Senghor, 1944), Efik (Ward, 1933) and Tiv (Arnott, 1958) display some form of vowel harmony in their use of affixes.

In some cases (Berry, 1955) there is said to be a distinction of voice quality between the sets of vowels involved in vowel harmony. I find it difficult to hear an auditory property which I can clearly assign as a distinguishing parameter of the two sets in any of these languages. The label tense/lax has been used simply for convenience, and should

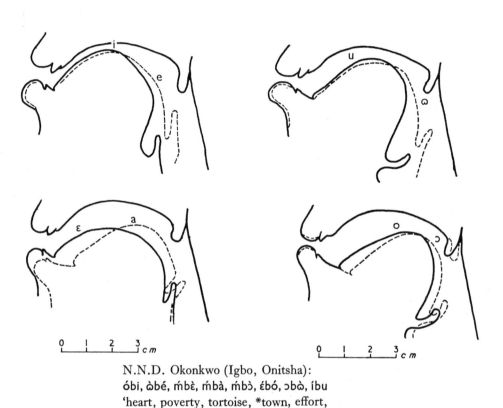

N.N.D. Okonkwo (Igbo, Onitsha):
óbi, ɔ̀bé, m̀bɛ̀, m̀bà, m̀bɔ̀, ɛ́bó, ɔbɔ̀, íbu
'heart, poverty, tortoise, *town, effort,
*person, it is, weight'   23 May 62

FIG. 12. Tracings from single frames in a cine-radiology film showing the tongue positions in the two sets of Igbo vowels.

not be interpreted as having a precise physiological meaning. The only data I have on the actual positions of the tongue in the articulation of the vowels of any of these languages are for Igbo. Figure 12 is based on tracings from a cine-radiology film taken during the pronunciation of the words listed in Table 8.

The positions shown are those in which the articulators approached nearest to a steady state in the second syllable of each word. The list of words cited was read twice; the measured difference between the positions of the articulators in a given vowel in the two readings was never as much as half the difference between that vowel and the adjacent vowel.

TABLE 8. Vowel contrasts in Igbo (Onitsha)

Vowel Harmony Set 1		Vowel Harmony Set 2	
óbi	heart	ọ̀bé	poverty of ability
ḿbɛ̀	tortoise	ḿbà	*town
ɛ́bó	*person	ḿbɔ́	effort
íbu	weight	ɔbọ̀	it is

Some of these pairs of vowels are very similar. There is little difference between i and e and o and ɔ from the point of view of the classic descriptions in terms of the highest point of the tongue. Nor, in these cases, is it any more profitable to use the more recent specification in terms of the cross-sectional area and place of maximum constriction of the vocal tract as suggested by Stevens and House (1955). The most striking difference between the vowels in the two sets is that in each case the body of the tongue is more

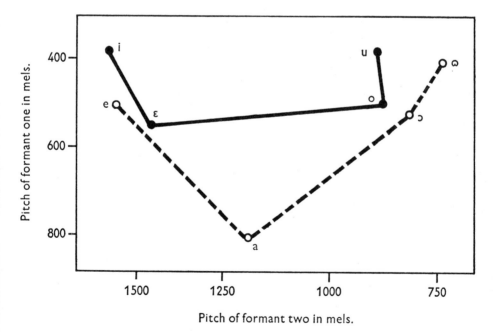

FIG. 13. Formant chart of the Igbo vowels shown in Table 8.

retracted for the vowels of Set 2. So it appears that there *is* a physiological parameter that distinguishes between these two sets of vowels, despite the fact that it is difficult to specify a unique auditory property that characterizes one or other set. Perhaps Bell (1867) had this property in mind when he described vowels not only in terms of the positions of the lips and the highest point of the tongue, but also with reference to the opening between the back of the mouth and the throat, which was usually 'primary' but could be enlarged so that it was 'wide'. Sweet (1906) may have been assessing something similar when he uses the terms 'narrow' and 'wide'. In the light of these X-ray studies of Igbo it is easier to make sense of his statement: 'In forming narrow vowels there is a feeling of tenseness in that part of the tongue where the sound is formed, the surface of the tongue being made more convex than in its natural "wide" shape, in which it is relaxed and flattened. . . The narrowing is the result not of raising the whole body of the tongue (with the help of the jaws), but of "bunching up" lengthways that part of it with which the sound is formed' (Sweet, 1906).

Some of the auditory characteristics of the two sets of Igbo vowels are shown in Figure 13. It may be seen that the vowels which have a greater constriction at the throat have higher formant frequencies (as would be expected). But the data of this figure alone do not indicate differences that could not be accounted for by specifications of the traditional auditory parameters which are labelled in terms of tongue height. It may be that we should retrain our ears, and either redefine the opposition tense/lax, or return to Sweet's narrow/wide terminology in order to make adequate descriptions of these vowels.

10

TONE AND INTONATION

There is little that need be said about tone and intonation from the phonetic point of view. They are, of course, very important aspects of West African languages. But the phonetic correlate – the rate of vibration of the vocal cords – is reasonably well understood, and is easily observable by instrumental means. An oscilloscope display of the fundamental frequency (Risberg, 1962), and narrow-band spectrograms which showed the harmonics, provided data for checking most of the tone markings in this monograph. (Many of these tone marks are, therefore, simply indications of what was said on a particular occasion; the accent notation should be regarded as impressionistic rather than phonological. The object of using a notation of this kind, before there had been an opportunity of discovering the lexical tone of all the forms cited, was simply to prevent the majority of cited forms from being completely ineffable.)

Certain phonetic aspects of tone occasionally cause investigators some difficulty. In West African languages (as in all others that I know of) some of the pitch variations are not linguistically significant, but are due to mechanical linkages which commonly (invariably?) occur in the process of speech production. These pitch variations are sometimes noted as allophones by investigators with sharp ears (Ansre, 1961). They are of two kinds, the one being an apparent effect of vowels on tones, and the other being associated with consonants. They can both be correlated with an economy in the physiological adjustments required in the production of speech.

The pitch of a sound depends on the tension of the vocal cords, which is largely determined by the intrinsic and extrinsic laryngeal musculature. All known linguistically significant variations in both tone and intonation languages are controlled through this musculature. But the tension of the vocal cords can also be affected by the action of other, more remote, muscles, which are used in articulatory movements. In pronouncing vowels such as i and u the whole of the tongue is raised up toward the roof of the mouth. As we noted earlier (page 5), the tongue is attached to the superior part of the hyoid bone, and some of the laryngeal muscles are attached to the inferior part. When the tongue is raised these laryngeal muscles are stretched, and the tension of the vocal cords is increased. Consequently, unless there is a deliberate slackening of the tension of the vocal cords, the pitch will increase whenever the tongue is moved up for an i or u vowel. We may therefore expect each linguistically significant tone to be slightly higher in syllables containing these vowels.

This hypothesis was tested by reference to Itsekiri. An informant pronounced a series of phrases of the form o fέ . . . mi 'he wants to . . . me' with the blank replaced by twenty-one different monosyllabic verb stems, one for each of the three tones which can occur on each of the seven vowel qualities. It was found that the high tone stems containing i and u had a peak fundamental frequency which was slightly more than 5 cps higher than the mean peak high tone frequency relative to this frame; similarly the fundamental frequency in the stems with mid and low tones was about 4 cps higher when the vowel was i or u. These differences are clearly audible, but they are slightly less than those found by Lehiste and Peterson (1961), who studied the effect of vowel and consonant quality on the fundamental frequency in English stressed monosyllables in the frame

'Say the word . . . again'. The discrepancy may not reflect a difference between English and Itsekiri. It is just as likely to be due to differences between the styles of speech of the informants, which may have been more emphatic or deliberate in the case of the English informant, and more colloquial in the case of the Itsekiri speaker. But in any case, the fact remains that there is in both languages an observable effect of vowel quality on pitch.

As Lehiste and Peterson also noted, there is often an observable correlation between consonants and pitch. There are two physiological reasons for this correlation. In the first place the raising of the tongue in palatal and velar consonants has an effect on the tension of the vocal cords similar to that of the high vowels. These consonants, therefore, tend to cause an increase in pitch. Secondly many consonant articulations cause a variation in the rate of flow of air through the glottis. This results in a variation in the pitch because the rate of vibration of the vocal cords depends in part on the force with which the cords are blown apart and sucked together by the airstream. Voiced stops and fricatives tend to cause a decrease in the rate of flow of air through the glottis, and hence a lowering of the pitch, since, in accordance with the principle of economy of effort, all the speakers I have observed do not bother to make the delicate adjustments in the tension of the vocal cords which would compensate for the decrease in flow. Conversely, during the first part of a vowel after the release of a voiceless stop or fricative there is a high rate of flow, which results in an increase in pitch as long as there is no counteracting adjustment in the tension of the vocal cords. It follows that the actual frequency of a given tone as pronounced by a given speaker will vary in accordance with the consonants at the beginning of the syllable. Ansre (1961) noted that in Ewe a given tone has a lower allotone in a syllable beginning with some voiced consonants. (Other investigators (Rycroft, 1960) have noted similar effects in other African languages.) It would seem that in Ewe (and probably in all other languages) there is a direct correlation between the linguistically significant tone and the action of the laryngeal musculature. Irrespective of the type of consonant onset or the type of vowel, the laryngeal musculature is adjusted in a given way for a given tone.

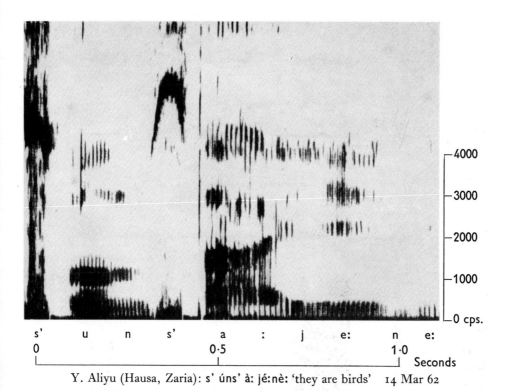

s'   u   n   s'   a   :   j   e: n   e:

0   0·5   1·0

Seconds

Y. Aliyu (Hausa, Zaria): s' úns' à: jé:nè: 'they are birds'   14 Mar 62

PLATE IA

A wide-band spectrogram illustrating Hausa ejective s' (see page 5).

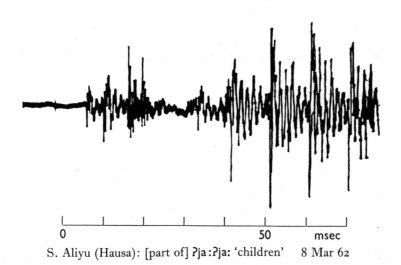

0   50   msec

S. Aliyu (Hausa): [part of] ʔja:ʔja: 'children'   8 Mar 62

PLATE IB

Waveform showing irregular vibrations of the vocal cords during an intervocalic
laryngealized approximant in Hausa (see page 16).

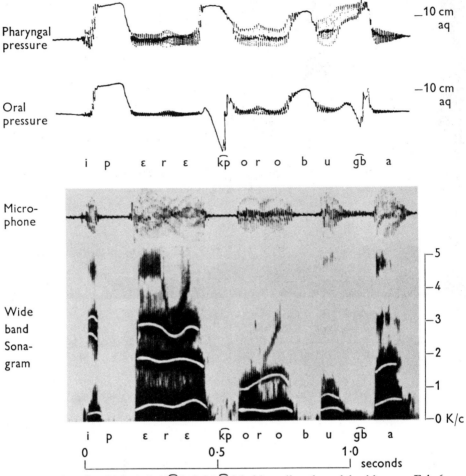

Pharyngal pressure   —10 cm aq

Oral pressure   —10 cm aq

i   p   ɛ   r   ɛ   k͡p   o   r   o   b   u   g͡b   a

Micro-phone

Wide band Sona-gram

—5

—4

—3

—2

—1

—0 K/c

i   p   ɛ   r   ɛ   k͡p   o   r   o   b   u   g͡b   a

0   0·5   1·0

seconds

A. Opubor (Itsekiri): ìpὲrέ k͡pòrò búg͡bá 'a big nail and a calabash'   19 Feb 62

PLATE 2A

The difference between p and k͡p and b and g͡b in Itsekiri (see page 9). Note also
the differences in aspiration (discussed on page 14).

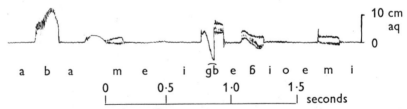

10 cm aq

0

a   b   a   m   e   i   g͡b   e   ɓ   i   o   e   m   i

0   0·5   1·0   1·5

seconds

M. T. Akobo (Kalabari): àbà mè íg͡bé ɓìò émì 'the fish is in the box'   2 May 62

PLATE 2B

The contrast between b, g͡b, ɓ in Kalabari as shown by variations in the recorded
mouth pressure (see page 10).

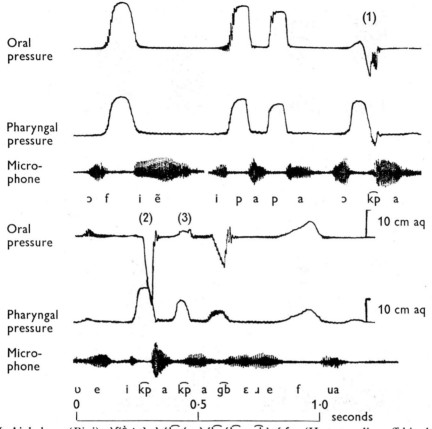

(1)

Oral
pressure

Pharyngal
pressure

Micro-
phone

ɔ f i ẽ   i p a p a   ɔ k͡p a

(2)   (3)

Oral
pressure

10 cm aq

Pharyngal
pressure

10 cm aq

Micro-
phone

ʊ e i k͡p a k͡p a g͡b ɛ ɹ e f ua
0   0·5   1·0
⌞_____⌞_____⌞ seconds

A. I. Aigbekaen (Bini): ɔ́fíẽ̀ ipàpà ɔ́k͡pá ʊè ík͡pák͡pa g͡bɛ̀ré fua 'He cut a slice off his skin'

8 Mar 62

PLATE 3A

Pressure records showing the variations in the production of k͡p in Bini (see page 9).

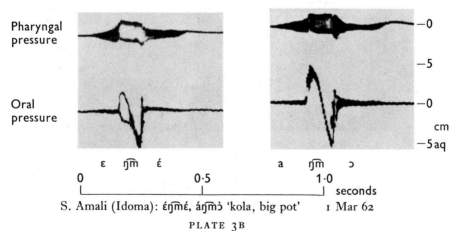

Pharyngal
pressure

Oral
pressure

—0

—5

—0

cm

—5aq

ɛ ŋ͡m έ   a ŋ͡m ɔ
0   0·5   1·0
⌞_____⌞_____⌞ seconds

S. Amali (Idoma): έŋ͡mέ, áŋ͡mɔ̀ 'kola, big pot'   1 Mar 62

PLATE 3B

Pressure records which show the use of a velaric airstream mechanism in
Idoma labial velar ŋ͡m (see page 12).

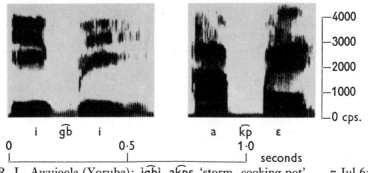

R. L. Awujọọla (Yoruba): ìg͡bì, ak͡pɛ 'storm, cooking pot'   7 Jul 62

Spectrograms of two Yoruba words, showing some of the acoustic characteristics of labial velar stops (see page 12).

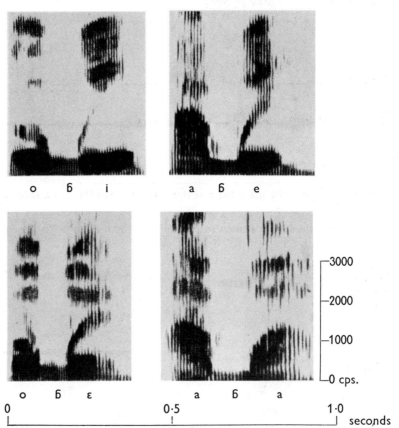

N. N. D. Okonkwo (Igbo, Onitsha): ɔ́ɓi, àɓè, ɔ́ɓɛ, àɓà 'dumb, eight days hence, poor, jaw' 23 May 62

Spectrograms illustrating the particular phonetic quality associated with Igbo implosive consonants (see page 13).

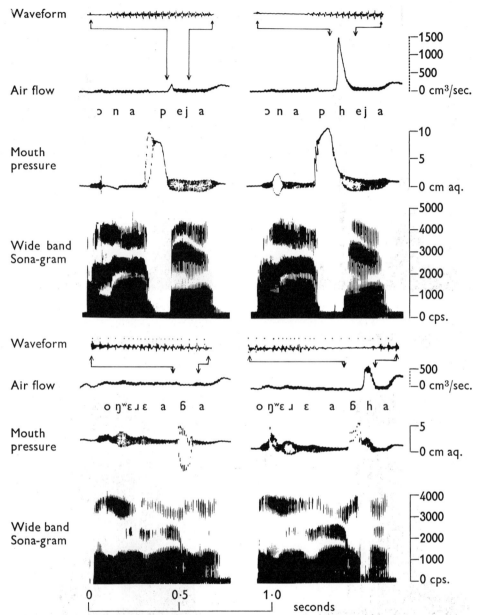

J. U. Madukwe (Igbo, Owerri): ɔ́nà àpé ja. ɔ́nà àphé jà 'He is pressing it. He is sharpening it' ó ŋwèɹè áɓà. ó ŋwèɹè áɓhà 'He has influence. He has a jaw'   5 May 62

PLATE 5
Data illustrating the difference between aspirated and unaspirated, voiced and voiceless stops in Igbo (dialect spoken in Owerri Province) (see page 14). The expanded waveform in the middle of the picture shows the approximate doubling of the rate of occurrence of glottal pulses which often occurs in laryngealized voicing (see page 16).

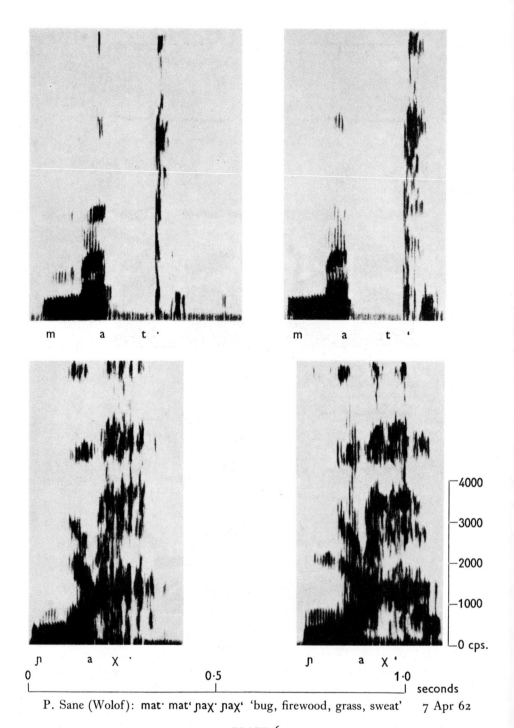

m a t ·

m a t '

ɲ a χ ·

ɲ a χ '

4000	3000	2000	1000	0 cps.

0   0·5   1·0   seconds

P. Sane (Wolof): mat· mat' ɲaχ· ɲaχ' 'bug, firewood, grass, sweat'   7 Apr 62

PLATE 6

Wide-band spectrograms illustrating the difference between whispered and voiceless final consonants in Wolof (see page 15).

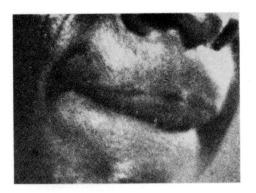 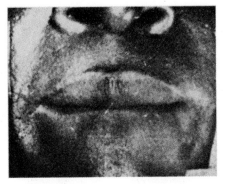

D. A. Sulemanu (Margi): bə́vᵇú (an adjective descriptive of sudden appearance
and flight)   10 May 62

Photographs of labiodental flaps in two different utterances in Margi. The left-hand
photograph was taken at a slightly earlier point in the sound than the right (see page 18).

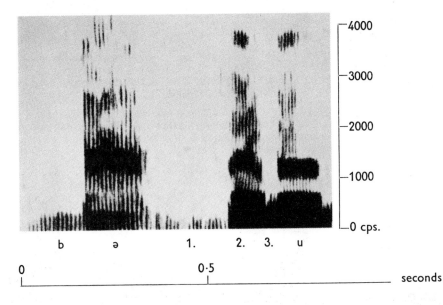

D. A. Sulemanu (Margi): bə́vᵇú (an adjective descriptive of sudden appearance
and flight)   10 May 62

A spectrogram illustrating a labiodental flap in Margi (see page 18).

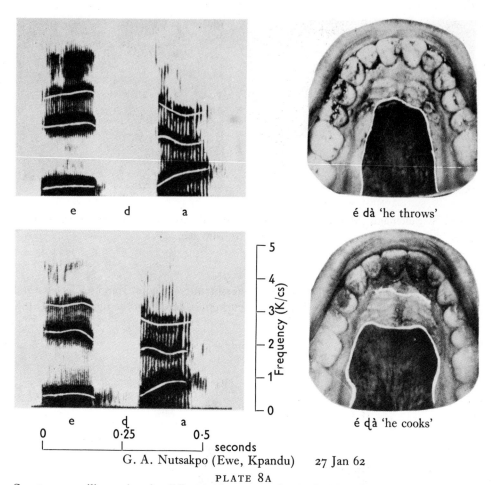

e d a   é ḍà 'he throws'

e ḍ a   é ḍà 'he cooks'

0   0·25   0·5   seconds

G. A. Nutsakpo (Ewe, Kpandu)   27 Jan 62

PLATE 8A

Spectrograms illustrating the difference between a (laminal) denti-alveolar and an (apical) post-alveolar voiced stop in Ewe (see page 20).

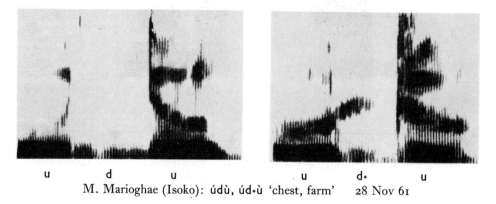

u   d   u   u   d̟   u

M. Marioghae (Isoko): údù, úd̟ù 'chest, farm'   28 Nov 61

PLATE 8B

Spectrograms illustrating the difference between an apical alveolar and a (laminal) denti-alveolar voiced stop in Isoko (see page 19).

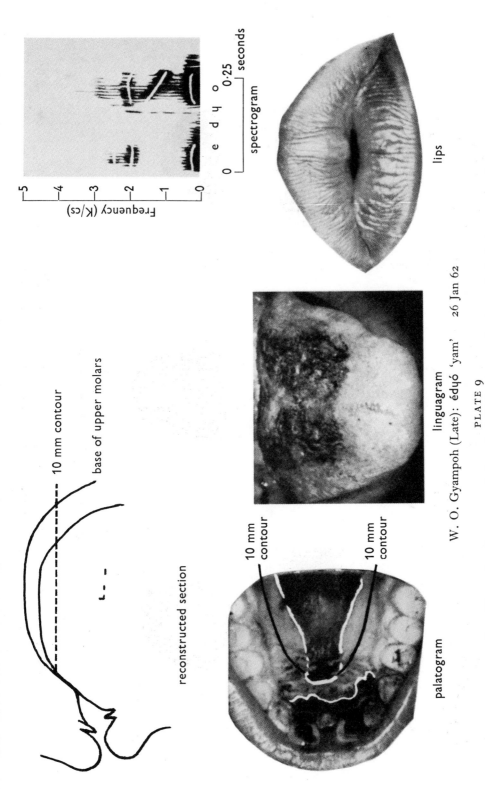

spectrogram

lips

linguagram

palatogram

reconstructed section

10 mm contour

base of upper molars

10 mm contour

10 mm contour

Frequency (K/cs)

seconds

e d ɥ o

W. O. Gyampoh (Late): édɥó 'yam'   26 Jan 62

PLATE 9

Data illustrating the pronunciation of a labialized pre-palatal stop in Late (see page 20).

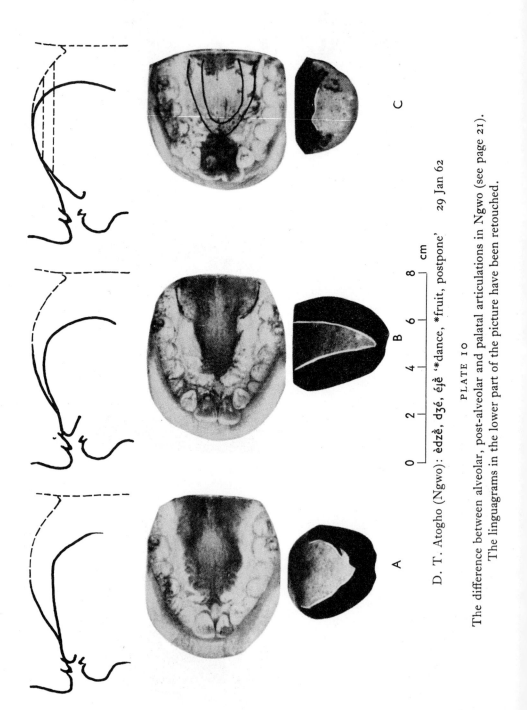

A   B   cm   C

PLATE 10

The difference between alveolar, post-alveolar and palatal articulations in Ngwo (see page 21).
The linguagrams in the lower part of the picture have been retouched.

D. T. Atogho (Ngwo): èdzɛ̀, dʒé, éjɛ̀ 'dance, *fruit, postpone'   29 Jan 62

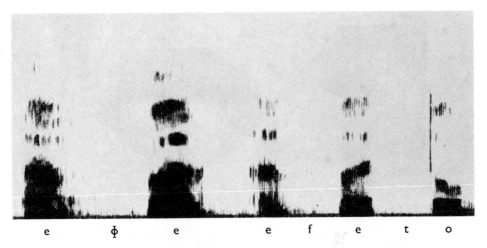

e   ɸ   e   e   f   e   t   o

A. Vowɔtɔ (Ewe): èɸé èfé tò 'His nails have grown'   29 Jun 62

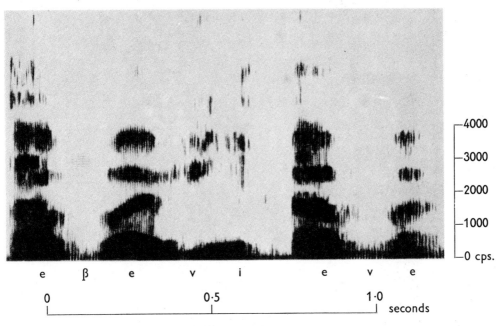

e   β   e   v   i   e   v   e

—4000
—3000
—2000
—1000
└—0 cps.

0   0·5   1·0

seconds

A. Vowɔtɔ (Ewe): èβè ví èvè 'Two Ewe children'   29 Jun 62

PLATE 11

Spectrograms illustrating the difference between voiced and voiceless bilabial and labiodental fricatives in Ewe (see page 25).

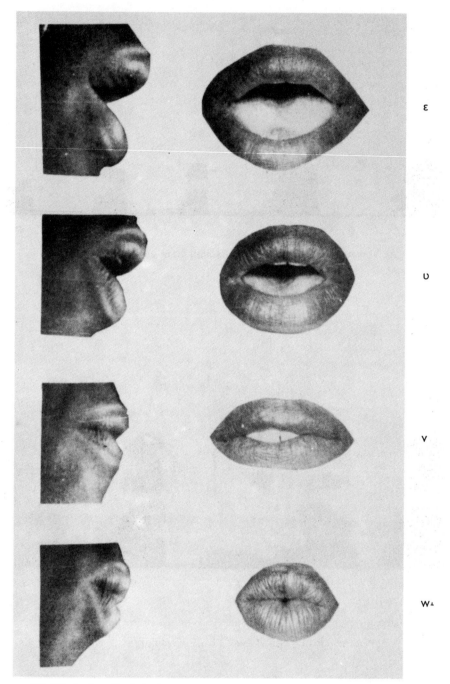

ɛ

ʊ

ʌ

wˑ

M. Marioghae (Isoko): ɛ́ʊɛ́, ɛ́vɛ́, ɛ́wˑɛ́ 'breath, how, hoe'   29 Nov 61

PLATE 12
Simultaneous full-face and side-view photographs showing the lip positions in Isoko ʊ ʌ wˑ
(see page 25).

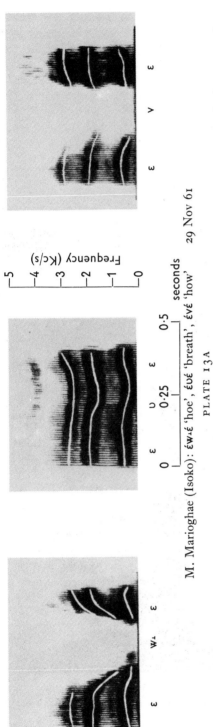

Frequency (Kc/s)

M. Marioghae (Isoko): ɛ́wʌɛ́ 'hoe', ɛ́uɛ́ 'breath', ɛ́vɛ́ 'how'

29 Nov 61

PLATE 13A

Spectrograms illustrating the difference between labial velar and labiodental fricatives and a bilabial approximant in Isoko (see page 25).

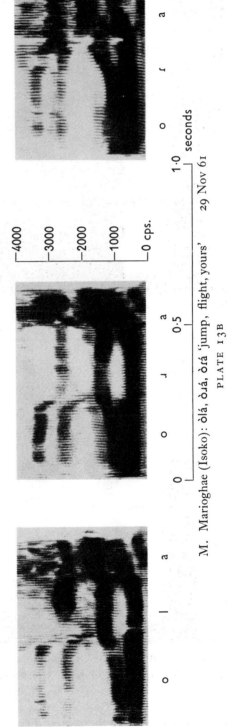

M. Marioghae (Isoko): òlá, òɹá, òɾá 'jump, flight, yours'

29 Nov 61

PLATE 13B

Spectrograms illustrating the difference between an alveolar lateral l, approximant ɹ, and tap ɾ in Isoko (see page 29).

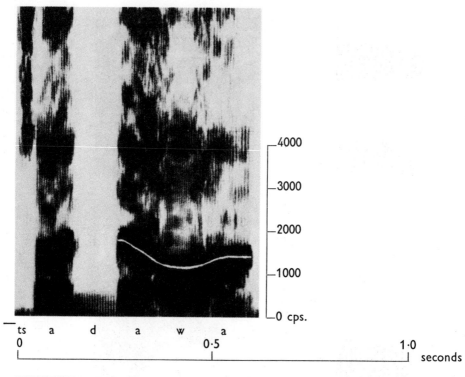

4000

3000

2000

1000

0 cps.

ts	a	d	a	w	a

0   0·5   1·0

seconds

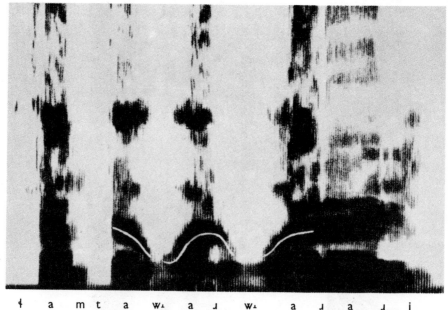

ɬ   a   m   t   a   wɪ   a   ɹ   wɪ   a   ɹ   a   ɹ   i

A. S. Mshelbwala (Bura): tsá dàwá. ɬámtà wɪaɹwɪáɹ ɹáɹì 'He is an enemy. Cut his throat'
23 Nov 61

PLATE 14

Spectrograms illustrating the contrast between labial velar fricatives and approximants
in Bura (see page 26). Note also the voiceless lateral (see page 29).

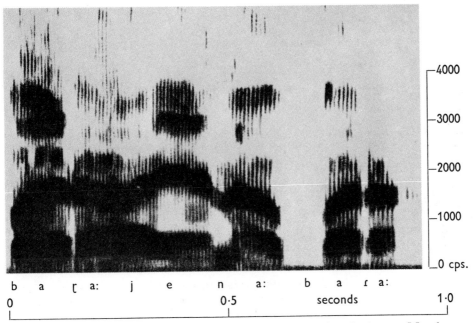

b a ɽ a: j e n a: b a ɾ a:

0   0·5   seconds   1·0

Y. Aliyu (Hausa, Zaria): báɽà: jénà: báɾà: 'The servant is begging'   14 Mar 62

PLATE 15A

Spectrogram of a Hausa utterance illustrating the difference between a flap ɽ and a tap ɾ
(see page 30).

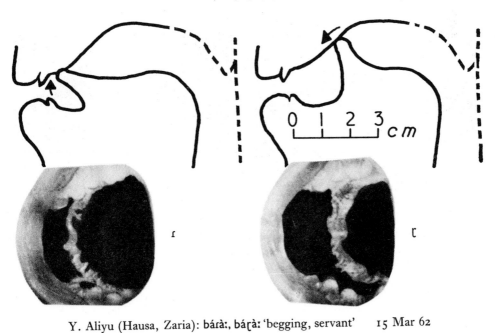

Y. Aliyu (Hausa, Zaria): báɾà:, báɽà: 'begging, servant'   15 Mar 62

PLATE 15B

Palatograms (retouched) and deduced tongue positions in Hausa ɾ and ɽ (see page 30).

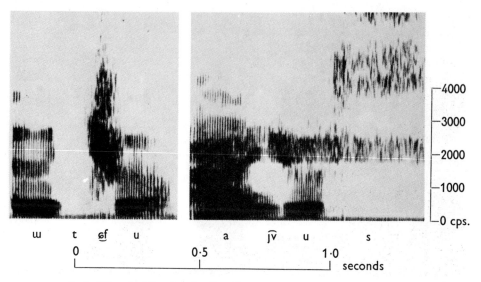

F. I. Nkwain (Kom): ɯ́tɕfù, aɟ͡v̄us 'mouth 'spirit,   29 Jan 62

PLATE 16A

Spectrograms of words containing labiodentalized articulations in Kom (see page 31).

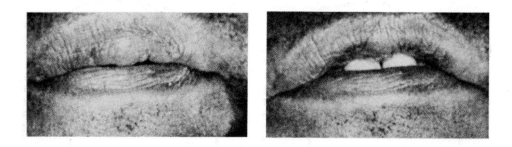

K. Avoke (Logba): ùβà – ùvá 'measles, side'   23 Jan 62

PLATE 16B

Photographs of the contrasting lip positions in bilabial and labiodental fricatives in Logba
(see page 25).

INFORMANTS AND LANGUAGES

The number before each language is the number on the map. The occupations of informants who were not students at a university or a teacher training college are given in brackets after the name. First names are given in full for all female informants. Only the chief informants for each language are listed.

NIGER-CONGO

WEST ATLANTIC

11	Fula	(a) Umaru (Herdsman), Mopti, Niger.
		(b) D. Laya, Say, Cercle de Niamey, Niger.
		(c) Y. Abubakar, Gombe, Bauchi Province, Nigeria.
		(d) A. Z. Ahmed, Yola, Adamawa Province, Nigeria.
12	Wolof	P. Sane, Dakar, Senegal.
13	Serer	M. Diouf, Sine, Senegal.
14	Dyola	A. de Badji (unemployed), Sindian, Cercle de Bignonia, Senegal.
15	Sherbro	A. T. Kainyek, Bonthe, Bonthe District, Sierra Leone.
16	Limba	E. Dumbuya, Tonko, Kambia District, Sierra Leone.
17	Temne	Alice Kamara, Makeni, Bombali District, Sierra Leone.
18	Kissi	F. G. S. Sogbandi, Sandaru, Kailahun District, Sierra Leone.

MANDE

20	Loko	A. Sisei, Gbobana, Bombali District, Sierra Leone.
21	Soso	S. A. S. Conteh, Kukuna, Kambia District, Sierra Leone.
22	Mende	D. A. B. Minah, Pujehun, Pujehun District, Sierra Leone.
23	Kono	S. M. Musa, Kaindordu, Kono District, Sierra Leone.
24	Bambara	L. Sogodogo, Bamako, Mali.

VOLTAIC

31	Sisala	L. D. Luri, Tumu, Upper Region, Ghana.
33	Dagbani	A. H. Yelzoli (Schoolteacher), Yendi, Northern Region, Ghana.
34	Batonu (Bariba)	B. Tamako (Steward), Ilesha, Borgu, Ilorin Province, Nigeria.

KWA

41	Cama (Ebrie)	J. T. Loba (Catechist), Abidjan, Ivory Coast.
42	Nzima	A. K. Jabialu, Half Assini, West Region, Ghana.
43	Fante	W. P. Brown-Orleans, Cape Coast, Central Region, Ghana.
44	Effutu	D. A. Akyeampong, Senya Beraku, Central Region, Ghana.
45	Ga (Gã)	E. K. Martey, Accra, Ghana.
46	Twi	D. K. Opare-Sem, Akropong, Akwapim, Eastern Region, Ghana.
47	Kyerepong	P. O. Asante, Abiriw, Akwapim, Eastern Region, Ghana.
48	Late	W. O. Gyampoh, Larteh, Eastern Region, Ghana.
49	Anum	E. Apea, Anum, Volta Region, Ghana.
50	Ewe	J. K. Agbeti, Peki, Volta Region, Ghana.
51	Nkonya	C. T. Amoah, Wurupong Nkonya, Volta Region, Ghana.

52 Avatime (Siya)   E. K. Anku-Tsede, Gbadzeme Avatime, Volta Region, Ghana.
53 Logba   K. Avoke, Logba Tota, Volta Region, Ghana.
54 Sekpele (Likpe)   H. K. Owusu, Likpemate, Volta Region, Ghana.
55 Siwu (Lolobi)   E. Y. Egblewogbe, Lolobi Huyeasem, Volta Region, Ghana.
56 Krachi   J. Owusu, Adomai, Volta Region, Ghana.
57 Gɛ̃   A. Moevi, Anecho, Togo.
58 Fɔ̃   R. Vignon, Porto Novo, Dahomey.
59 Yoruba   R. L. Awujoola (Schoolteacher), Oyo, Oyo Province, Nigeria.
60 Bini   A. I. Aigbekaen, Benin City, Benin Province, Nigeria.
61 Itsekiri   A. Opubor, Elume, Warri, Delta Province, Nigeria.
62 Urhobo   G. E. Umokoro, Okpara, Delta Province, Nigeria.
63 Isoko   M. A. Marioghae, Aviara, Delta Province, Nigeria.
64 Ora   A. Ilevbare, Sabon Gida, Benin Province, Nigeria.
65 Igbira   A. Ozigi, Okene, Kabba Province, Nigeria.
66 Nupe   M. S. Angulu, Bida, Niger Province, Nigeria.
67 Igbo   J. U. Madukwe, Aja, Okigwi, Owerri Province, Nigeria.
68 Idoma   E. Obe, Orukpa, Idoma Division, Nigeria.
71 Ijaw (Upper)   C. B. Ndiomu, Odoni, Delta Province, Nigeria.
72 Kalabari   S. D. Wokoma, Buguma, Rivers Province, Nigeria.

BENUE-CONGO
80 Kambari   A. Kamashi (Farmer), Salka, Niger Province, Nigeria.
81 Ibibio   E. Ufot, Abak, Calabar Province, Nigeria.
82 Efik   J. Ekong, Uyo, Calabar Province, Nigeria.
83 Tiv   E. A. Akiga, Mkar, Benue Province, Nigeria.
84 Birom   C. Lodam, Jos, Plateau Province, Nigeria.
85 Kutep   C. A. Iyaba, Takum, Benue Province, Nigeria.
86 Ngwe   B. Foretia (Librarian), Fontem, Mamfe, Cameroun.
87 Ngwo   D. T. Atogho, Widekum, Mamfe, Cameroun.
88 Bafut   S. N. Shu, Bafut, Bamenda, Cameroun.
89 Kom   F. I. W. Nkwain, Njinikom, Wum, Cameroun.

AFRO-ASIATIC
CHAD
91 Hausa   (a) Y. Aliyu, Zaria, Zaria Province, Nigeria.
(b) S. Dawaki, Kano, Kano Province, Nigeria.
(c) S. A. Aliyu, Katsina, Katsina Province, Nigeria.
92 Bura   A. S. Mshelbwala, Biu, Bornu Province, Nigeria.
93 Margi   A. D. Sulemanu (Soldier), Duhu, Adamawa Province, Nigeria.

NILO-SAHARAN
95 Songay   D. Laya, Say, Cercle de Niamey, Niger.

KRIO
96 Krio   A. Cole, Waterloo, Colony, Sierra Leone.

This appendix is designed to serve as a comprehensive source of examples of the contrasts between the different consonantal features which occur in the West African languages I have been able to study. The lists contain an example of most of the consonant phonemes and, where appropriate, some of the consonant clusters in each of the languages shown on the map except for Kono (mentioned on page 11), Cama (page 14), Nupe (page 25), Ibibio (page 9), Birom (pages 1 and 31) and Ngwe (page 33). I have worked with informants of all these languages; but I do not have data which enable me to illustrate the phonetic nature of the consonant oppositions.

In the case of each of the other languages, the words are listed so that some of the consonants which sound similar occur next to each other; but as an over-riding condition to facilitate reference and comparison between languages, the contrasting phonetic elements which are illustrated are always given in the following order: (1) voiceless stops and affricates; (2) voiced stops and affricates; (3) nasals; (4) voiceless labial fricatives and consonantal vocoids; (5) voiced labial fricatives and consonantal vocoids; (6) other voiceless central fricatives and h; (7) other voiced central fricatives; (8) voiceless laterals and ɬ; (9) voiced laterals, trills, taps and lingual consonantal vocoids.

As we noted in the Introduction, the glosses are intended to serve simply as identifications of the cited forms. An asterisk denotes that the word is a proper name (*girl), or a particular item of the general kind (*fruit, *fish) for which I was unable to find a short English equivalent. Since the transcription is in a comparative phonemic style, positional variants (allophones) such as nasalized vowels following nasal consonants are not shown.

FULA (11)

Phonological contrasts suggested and elicited by D. Arnott (see also Labouret (1952), Wolff (1959) for different analyses). The doubled vowels could be written as single vowels plus length marks.

p	to paaʔdaa	Where are you off to?	g	o gooŋʔdii	he told the truth
t	o tawii·	he found	ŋg	ʔbe ŋgarii	they came
tʂ	ʔbe tʂoodii	they sold	m	o mahii	he built
k	ʔbe kaarii	they are replete	n	o naatii	he entered
ʔ	o ʔaawii	he sowed	ɲ	o ɲaamii	he ate
b	o barkiʔdii	he is fortunate	ŋ	o ŋatii	he bit
mb	no mbaalduʔdaa	Have you had a good night?	f	o faʔbbii	he delayed
ʔb	o ʔbamtii	he lifted	w	o warii[1]	he came
d	o dawii	he started early	v	o vami[2]	he wove
nd	o ndaarii	he looked	s	o soodii[3]	he bought
ʔd	o ʔdaanike	he slept	ʃ	o ʃaawii[2]	they wrapped
dʐ	o dʐalii	he laughed	h	o haarii	he is replete
ndʐ	toje ndʐaanoʔdaa	Where had you gone?	j	o jarii	he drank
			ʔj	o ʔjakkii	he chewed
			l	o loonii	he washed
			r	o remii	he hoed

[1] Mopti Fula has ʊ for w in some contexts.
[2] In Adamawa Fula; not in all dialects.
[3] Not in Adamawa Fula.

WOLOF (12)

Phonology according to Sauvageot (personal communication). Compare also Ward (1939a).

p	paːχ	hole	m	maːm	grandparent
t	tɛn	well	n	naːn	drink
tɕ	tɕaq	necklace	ɲ	ɲaq	sweat
k	buki	hyena	ŋ	ŋaːŋ	open the mouth
q	buqi	stare	f	fas	horse
b	baːχ	good	s	samm	shepherd
d	deːmba	yesterday	l	lɛm	bend
dʒ	dʒar	pass	r	raχas	wash
g	gan	host			

SERER (13)

Contrasts suggested by informant. Compare Greffier (1901).

p	paχ	wealth	g	gamb	[meaning obscure]
t	tig	thing	ŋg	ŋgelem	camel
c	cit	gift	m	maːm	grandparent
k	kid	eyes	n	neʔ	appoint
q	qos	leg	ɲ	ɲam	eat
b	bab	father	ŋ	ŋas	play
ʔb	ʔbaːχ	axe	f	fiʔ	make
mb	mbod	pollen	ʋ	ʋag	be able
d	dig	run	s	soʔb	find
ʔd	ʔdat	way	χ	χol	clean
nd	ndig	dry season	h	hiran	start late
ɟ	ɟir	illness	j	jaj	mother
ʔɟ	ʔɟal	gap in teeth	l	laj	say
ɲɟ	ɲɟag	*person	r	ret	go

DYOLA (14)

Phonology based on Weiss (1939).

p	bapalai	intimacy	n	kanap	wick
t	amata	shepherd	ɲ	kaɲalo	he chews
c	cok	chance	ŋ	eɟaŋa	girl
k	kafat	bar	ʋ	eʋel	make a noise
b	bala	before	f	faŋk	residence
d	edaba	trousers	s	basap	thief
ɟ	ɟat	today	j	ajala	magician
g	bagam	lawsuit	l	elag	repay
m	mafas	hay	ɹ	aɹana	drinker

46

APPENDIX B

SHERBRO (15)

Phonological contrasts elicited from the informant. Compare Sumner (1921).

p	pak	bone	n	nà	spider
mp	mpanθ	labour	ɲ	ɲanɔ́	stranger
t	tàmbàsé	sign	ŋ	ŋus	push
tʃ	tʃal	harnessed antelope	f	faká	village
			w	wántàmà	girl
ntʃ	ntʃaŋ	teeth	v	véi	thorn
k	káŋ	read	θ	θàm	grandmother
ŋkw	ŋkwέi	palm oil	nθ	nθé	cheeks
b	bà	father	s	sa¹	red
mb	mbàŋk	beads	h	hál	river
nd	ndɔ	where	ŋh	ŋhɔ́	talk
dʒ	dʒàdʒὲl	mother-in-law	j	ja	mother
ndʒ	ndʒó	food	l	lá	wife
g͡b	g͡bam	potato	r	rà	three
ŋ͡mg͡b	ŋ͡mg͡baŋ͡mg͡bàŋ	rib	ŋr	ŋrɔ́m	medicine
m	mam	laugh			

¹ In other dialects such as Shenge there is s–ʃ; sém 'stand', ʃa 'red'.

LIMBA (16)

Contrasts elicited from the informant with the aid of Mrs. Jo Roberts.

p	pama	scaring birds	g͡b	g͡bada	shade
mp	mpaṭi	children	ŋ͡mg͡b	ŋ͡mg͡bindi	sack
t̪	t̪eni	fowl	m	manaŋ	cow
nt̪	nt̪aŋka	cassava leaf	n	nama	how
tˢ	tˢɛni	medicine	ŋ	ŋajɛŋ	wood
nt̪ˢ	nt̪ˢant̪ˢa	anus	w	waṭe	man
k	kama¹	dance	s	sɛṭɛ	crowd
ŋk	ŋkala	*rope	h	hɛmõ²	old man
b	bara	meat	j	jɛrɛmɛ	woman
mb	mboro	history	l	loŋa	clothes
d	dɛŋka	ridge	r	raɣa¹	stone
nd	ndaŋka	*rope			

¹ k occurs word initial and after ŋ; g or ɣ occur elsewhere. They are allophones of the same phoneme.
² f occurs as a free variant of h in some words (but not in this one).

TEMNE (17)

Phonology as in Sumner (1922), with the addition of kʷ, ɲ, ʃ, ɥ: the phonological status of ɥ is not clear.

p	pét̪	town	ŋ	ŋàs	row
t̪	t̪or	descend	f	fàe	kill
tˢ	tˢor	farms	ɥ	ɥìr	goat
k	kɔ̀r	farm	w	wát̪	child
kʷ	kʷέ	why	s	sàmpà	dancer
b	bàná	banana	ʃ	ʃék	teeth
g͡b	g͡bàsá	head tie	h	hálì	though
d	dìr	mortar	j	jà	mother
m	mĕt̪ˢ	water	l	làs	ugly
n	nàk	rice	r	rès	mate
ɲ	ɲá	where			

47

KISSI (18)

Examples suggested by the informant. Compare Berry (1959).

p	poː	be equal	ɲ	ɲású	scratch
k͡p	k͡pákɔ	back	f	faŋgafáŋga	weighty
t	tófa	look	w	waláa	mat
k	kól	drink	v	vì	die
b	bándei	farm house	s	sóː	talk
d	de	mother	h	hakéjo	sin
m	ména	swear	j	jaláa	hammock
n	númbɔ̀	[interrogative]	l	límí	tell

LOKO (20)

Phonological contrasts elicited from the informant.

p	pàŋʷa	lizard	n	nàbénà	look there
k͡p	k͡pándì	boil it	ɲ	ɲahá	woman
tˢ	tˢàmbai	*drum	ŋ	ŋaŋgena	scratch it
c	cani	bottle	f	fã	whip
k	kánà	dance it	w	waà	come in
b	ba	come	s	sándé	*head tie
mb	mbà	rice	h	hábà	key
nd	ndaga	leaf	j	jaà	talk
ɲɟ	ɲɟà	water	l	laà	lie down
ŋg	ŋgápɹènà	I can do it	ɹ	ɹúánàmà	admit it
m	máɹònà	increase it			

SOSO (21)

Phonological contrasts elicited from informant.

p	pápù	many	ɲ	ɲaχón	sweet
k͡p	k͡pàkú	hang	ŋ	ŋasón	snatch
t	taɹa	brother	w	wálí	work
k	kakun	yawn	f	fa	come
b	babá	father	s	sáɹé	bed
g͡b	g͡bag͡bá	rice store	χ	χaχon	*tree
d	daŋgí	pass	h	hámɛ	worry
g	gasí	fine	j	ja	eye
m	màŋgé.	chief	l	làlà	paddle
n	nansé	what	ɹ	ɹafa	bring

APPENDIX B

MENDE (22)

Phonology in accordance with Crosby and Ward (1944), with additional suggestions by G. Goba.

p	hàpíí	root	m	mànîi	palm fibre
k͡p	hák͡péí	sauce	n	náánì	four
t	pààtèí	marsh	ɲ	ɲààpùí	woman
k	mbàkèí	organ	ŋ	ŋàɲìí	sand
b	fáábòì	parrot	f	fádʒíí	bucket
g͡b	hàg͡bìí	rabbit	w	háwéí	laziness
d	màmádà	grandfather	v	kàvéi	big handful
dʒ	kàdʒèí	plassava	s	sókúí	corner
mb	mbóg͡bèí	cutlass	h	háwéí	laziness
nd	ndɔ̀ɔmìí	ground	j	kàjìí	rust
ndʒ	ndʒèpèí	talk	l	lɔ́ɔlù	five
ŋg	ŋgàféí	devil			

BAMBARA (24)

Contrasts largely as indicated in Bazin (1906).

p	pàrà	pace	n	naa	belly
t	ta	possession	ɲ	ɲàgá	nest
tɕ	tɕàgà	prostitute	ŋ	ŋànìá	malice
k	kã	voice	f	fà	father
b	bá	mother	w	wàrà	beast
d	da	opening	s	sá	snake
dʒ	dʒá	shadow	h	háké	right
g	gà	channel	z	zánfàlà	[meaning obscure]
gʷ	gʷá	family	j	jàfá	pardon
m	mɔ̀gɔ́	mankind	l	lábàn	end

ɣ and g, and possibly ɹ/ɾ and l appear to be in free variation intervocalically.

SISALA (31)

Contrasts suggested by informant. In the frame mì . . . nè 'It is my . . .' the tones were always high for all items.

p	pire	hoe	ɲ	ɲile	horn
k͡p	k͡paha	chair	ŋ	ŋaliiŋ	drawing water
t	tasa	bowl	f	foli	white man
cɕ	cɕaŋ	broom	w	wia	sun
k	kaha	thatch	v	vaha	dog
b	baga	farm	s	saŋ	axe
g͡b	g͡baŋa	calabash	h	hɛnɛ	pot
d	daŋ	stick	z	zara	*bird
ɟʒ	ɟʒᵃba	horse	j	jɔbo	market
g	gara	thief	l	lorimiŋ	louse
m	mami	frock	r	pire	hoe
n	naŋ	foot			

49

DAGBANI (33)

Phonological contrasts suggested and elicited by A. Wilson.

p	opahia	he cleared the ground	ɲ	oɲamja	he ground
k͡p	ok͡pa:ja	he poured	ŋ	oŋahja	he harvested corn
t	otamja	he forgot	f	ofa:ja	he let go
t∫	ot∫aŋja	he went	w	owaja	he danced
k	okaria	he drove away	v	ovalma	he alerted
b	obaŋia	he knew	s	osaja	he planted
g͡b	og͡ba:ja	he got hold	∫	o∫ɛja	he sewed
d	odamja	he shook	z	ozaŋja	he put
d̪ʒ	od̪ʒaŋa	his monkey	ʒ	oʒɛja	he blew
g	ogaria	he passed	j	oja:ja	he took from water
m	omala	he had	ɣ	opaɣia	he washed
ŋ͡m	oɲm̃aja	he mashed	l	olabja	he returned
n	ona:ja	he finished			

BATONU (34)

Contrasts suggested by informant. This language is related to Bariba (cf. Welmers 1952 for a very similar analysis).

p	ápàlì	you shine	m	ámɔ́	you have
k͡p	ák͡pá	you finished	n	ànà	you come
t	átɛ̀	you are late	f	fàfàrú	fan
k	ákàdà	you go with it	w	áwá	you see
b	àbèrú	jacket	s	ásá	head tie
g͡b	ág͡bédà	you are going to the farm	j	ájàrà	you go out
			l	wɔlɔ́	top
d	ádà	you go	r	wɔ́rú	hole
g	ágó	you kill it			

NZIMA (42)

Phonological contrasts elicited from informant (cf. Berry 1955).

t͡p	ót͡pì[1]	it is thick	ŋ	jèŋélèbè	our wisdom
k͡p	jek͡pólè[2]	big	ŋʷ	ɛŋʷìà	sand
t	otí	his head	f	jɩfélɛ́	his fish
cɕ	ɔcɕɛ̀	he divides	ɥ	óɥìè[1]	he has finished
cɕʷ	ɔcɕʷɛ̀	he pulls	w	èwɔ̀lɛ̀[2]	a snake
k	òkìlè	he teaches	v	ɛvɛlè	fatigue
kʷ	ŋkʷáné	soup	s	jɩsɛ́lɛ́	his parent
b	óbíà	he washes	∫	jé∫è	he has caught
d	ódì	he eats	h	jɩhálɛ́	his sore
ɟ	oɟé	his teeth	z	èzìlé	recognition
ɟɥ	oɟɥè	it irritates	j	éjèlè	cold
m	jɛmélɛ́	our palm trees	ɣ	èɣàlɛ́	coming
n	ɛnɛ́	today	l	èlàlɛ́	dream
ɲ	jèɲélìɛ̀	our behaviour			

[1] occurs before front vowels only.

[2] occurs before back vowels only. t͡p and k͡p, ɥ and w, are allophones of the same phoneme.

FANTE (43)

Phonology based on Welmers (1946).

p	pàpá	father	m	ɔ́mà	he gives
t	atá	twin	n	ɔ̀nàm	he walks
ʦ	ɔ́ʦè	he catches	ɲ	ópà	he obtains
ʦʷ	óʦʷà	he cuts	f	ɔ́fà	he takes
k	ɔkà	he bites	ɥ	ɔ̀áɥè	he chewed it
kʷ	kʷatá	leprosy	w	ɔ̀sáwèè	he danced
b	óbù	he reaps	s	ɔ́sà	he syringes
d	ɔ́dà	lies	ɕʷ	ɔ́ɕʷɛ̀	he looks at
ʥ	ódʑà	he leaves	h	ɔ́hàm	he quarrels
ʥʷ	ɔ́dʑʷè	he calms	j	ɔ́jàm	he grinds
g	àgór	play	r	àrà	[adverb]

EFFUTU (44)

Phonological contrasts elicited from the informant.

k͡p	k͡pák͡pa[1]	good	g	ìgó	house
t̪	àt̪ɛ́	Dad	m	émà	dirt
ʦ	ɔ́ʦe	tomorrow	n	àná	leg
ʦʷ	eʦʷér	ladder	ɲ	òɲí[3]	male
k	ekà	pad	f	éfʊ̀	wind
b	àbá	hand	w	ɔ̀wó	breast
g͡b	àg͡bébí[2]	animal	s	ɔ̀sá	sponge
d̪	àd̪á	song	h	áha	guinea worm
ʥ	ídʑà	fire	j	ája	Mum
d̪ʷ	ídʑʷò	yam	l	ɔ̀làfá	hundred

[1] k͡p probably p velarized before back vowels, and p before front vowels.
[2] g͡b probably b velarized in this, the only word known to me with g͡b.
[3] ɲ probably is an allophone of j in nasalized syllables.

GÃ (45)

Phonology in accordance with Berry (1951a).

p	apáŋ	meeting	n	ènà	he saw
k͡p	ék͡pá	he stopped	ɲ	èɲá	he rejoiced
t	étè	he went	ŋ	èŋà	his wife
ʦ	éʦò	it is ablaze	f	èfà	he lent
ʦʷ	eʦʷa	he struck	ɥ	èɥérélà	he sat by the fire
k	éka	it struck	w	èwè	his house
b	èbà	he came	s	esa	it is fitting
g͡b	egbo	he is dead	ɕ	éɕá	sin
d	édɔ̀	it is bent	ɕʷ	éɕʷá	it is scattered
ʥ	èdʑà	it is right	h	ehá̃	he gave
ʥʷ	edʑʷa	it is broken	z	àzìí	spell
g	égɔ̀ì	he belched	j	òjá	quickly
m	èmà	he borrowed	l	mlà	law
ŋ͡m	èŋ͡mɔ̀	he laughed	ɻ	mɻá	quickly

TWI (46)

Phonology based on Christaller (1881) and Ward (1939b) with additional suggestions by Stewart (personal communication).

p	ɔ́pà	he chooses	n	ɛ̀ná	mother
t	àtá	twin brother	ɲ	óɲàm	he is waiting
c	càcà	straw mattress	f	ɔ́fà	he takes
cɥ	ócɥà	he cuts	ɥ	wàáɥè	he chewed it
k	ɔ́kà	he bites	w	ɔsáwèl	he danced
kʷ	ŋkʷá	life	s	ɔ́sà	he syringes
b	àbá	fruit	ɕ	ɔ̀ɕé	boundary
d	ɔ́dà	he lies	ʃʷ	ɔ́ʃʷɛ̀	he looked at
ɟ	ɛ̀ɟá	father	h	ɛ̀húm	tornado
ɟɥ	óɟɥò	he is gentle	j	ɔ́jɛ̀	he does
g	àgóɹú	play	ɹ	àɹà	[adverb]
m	ɔ́mà	he gives			

KYEREPONG (47)

Phonological contrasts elicited from the informant.

k͡p	akp̄à[1]	he is tall	n	ɔ́ná[3]	leg
t	ɔtá	hatred	ɲ	óɲà	wringing
tɕ	atɕɔ̀	mattock	f	áfe	rope
k	àkɛ́	boundary	w	ɔwɔ́	snake
b	àbàá	stick	s	ásà[3]	broom
d	àdɛ́	cutlass	ʃ	ɛʃé[2]	sap
dɥ	ódɥé[2]	yam	h	àhɛ́	noon
ɟ	ɛ́ɟl	fire	j	àjlɛ́	he says
m	áml	stomach	l	álɛ́[3]	water pot

[1] k͡p probably velarized p.
[2] dɥ is rare.
[3] often velarized.

LATE (48)

Phonological contrasts elicited from informant.

p	ɔ́pɛ́	harmattan	m	àmɛ́	stomach
k͡p	akp̄o	mortar	n	ánò	mouth
t	éta	cloth	ŋj	éŋjɛ̀	night
tɕ	étɕà	building	ʍ	éʍè	Saturday
tɕʷ	ɛ̀tɕʷó	*disease	f	áfè	axe
k	ékɔ̌	fight	w	àwe	calabash
kʷ	ákʷàá	joints	s	ɛ̀só	ear
b	ébɛ̀	palm nut	ç	éçèɹè	shyness
d	àdɛ́	cutlass	j	ójì	tree
dʐ	ɛ̀dʐáŋ	spear	l	òló	sore
dɥ	édɥò	yam	ɹ	áɹá	[modifier]
g	agò	silk			

ANUM (49)

Phonological contrasts elicited from informant.

k͡p	àk͡pέ	path	n	ɔná	foot
t̪	t̪ɪέt̪ɪ	long	ɲj	ὲɲjé	man
tɕ	àtɕέ	tomorrow	ŋʷ	àŋʷá	nose
tɕʷ	έtɕʷùkʷɪ̀	he is pulling out	ʍ	ámὲ	new
k	ákókʷè	she has given birth	f	àfὲt̪ὲ	moon
kʷ	ékʷà	neck	w	ewéèbì	star
b	àbóbì	bird	s	ɔsó	ear
d	àdàmé	heart	ʃ	ὲtɔʃîì	goat
dʑ	ɛdʑési	smoke	h	àhó	breast
dʑʷ	ὲdʑʷàéhɔ̀	proud person	j	έjì	tree
gʷ	έgʷà	he has run away	l	jèlí	stand
m	àdàmé	heart			

EWE (50)

Phonology and examples according to Ansre (1961). Compare Westermann (1930) and Berry (1951b).

p	épɔ́	he was wet	ɲ	éɲa	he knew
k͡p	ek͡pá	he faded	ŋ	éŋé	he broke
t	étá	he crawled	ɸ	éɸá	he polished
ts	étsɔ̀	he cut	f	éfá	he was cold
tɕ	ètɕó	canine tooth	w	éwɔ̀	he made
k	éká	he chipped	β	ὲβὲ	the Ewe language
b	éba	he cheated	v	ὲvὲ	two
g͡b	ég͡bá	he roofed	s	ésá	he douched
d	édà	he throws	x	éxé	he paid
ɖ	éɖà	she laid an egg	h	éhé	he disappeared
dʑ	édʑá	he cut vigorously	z	ézá	he used
g	égǎ	he became well off	j	éjá	it went bad
m	émá	he divided	ɣ	éɣá	he scraped
n	éná	he gave	l	élé	he caught

Clusters (in which r and l are in complementary distribution and could be regarded phonemically as l).

pl	éplá	he hurt	ŋl	éŋlɔ́	he wrinkled
k͡pl	ék͡plá	he girded	ɸl	éɸlè	he bought
tr	étrɔ́	he turned	fl	éflé	he split off
tɕr	étɕrì	he hated	wl	éwlɔ́	he hid
kl	éklɔ́	he washed	βl	ὲβlɔ́	mushroom
bl	éblá	he bound	vl	évló	he is evil
g͡bl	àg͡blè	farm	sr	ésrɔ́	he studies
dr	édró	he put down	xl	éxlɔ́	he advised
dʑr	édʑré	he quarrelled	hl	ehlɔ	it is bruised
gl	églɔ́	it is bent	jr	éjrǎ	he blessed
ml	émlǎ	he tamed	ɣl	éɣlá	he hid
ɲr	éɲrǎ	he sharpened			

NKONYA (51)

Phonological contrasts elicited from informant.

p	ɔ́pa	noon	g	àgò[2]	velvet
k͡p	ɔ́k͡pa	way	m	ɔmá	nation
t̪	t̪at̪i	cloth	n	ana	four
ts	ɔ́ɔ́tsà	he will dance	ɲ	aáɲá	he has got
t̪ɥ	át̪ɥɛ̃	*game	ŋ	ɔŋɔ́té	nose
c	ɔ́ɔ́cè	he will look	ʍ	ahoʍɛ	mirror
k	ɔka	wife	f	afá	drug
kʷ	àkʷàdú	bananas	w	ɔ́ɔ́wà	he will dress
b	ɔbá	he is coming	s	àsà	three
d	òódù	he will bite	j	ɔ́yɔ	he is going
dʑ	odʑá	fire	l	ɔ́lɔ̀	water pot
d̪ɥ	id̪ɥó	yam	ɹ	ɔsuɹò	land

[1] k͡p probably velarized p.
[2] g is rare.

AVATIME (SIYA) (52)

Phonological contrasts elicited from informant.

p	ápè	she is beautiful	ŋ	áŋà	he has eaten
k͡p	ák͡pé	it is worn	ʍˑ	áʍˑá	charcoal
t	átà	he has chewed	f	áàfume	I will be fed up
ts	átsá	he has paid	w	áwà	he has done . . .
k	ákɔ̀	he has taken	β	áβà	on top
b	ábá	he has come	v	ávè	*fruit
g͡b	ág͡bà	he is paralyzed	s	ásá	he has rung
d	édú	he installs	h	áhà	he has collected
dz	ádzé	he has been there			sticks
g	ágú	he has plucked	z	ázɛ	he has got
m	áàme	he will tap (palm wine)	j	ájá	he has cut
n	ána	he will hammer	l	álɛ	he stood

LOGBA (53)

Phonological contrasts elicited from the informant.

p	ùpùé[1]	*cassava	n	ìnɔ́	meat
k͡p	ák͡pá	foot	ɲ	ɔ́ɲɛ́ɛ́	he's caught him
t	até	he says	ŋ	ɔŋɔɲi	he writes
ts	atsá	weariness	ɸ	sɛɸoɸo	flower
c	ɔ́cámì	chief's spokesman	f	ɔ́fà	uncle
k	úkú	bone	β	ùβà	measles
b	abíá	stool	v	ùvá	side
g͡b	ág͡bɔ̀	tumour	s	ɔsá	man
d	ɔ́dɛɛ	he scolds him	ʃ	oʃibíe	he cuts him
dz	ɔdzá	fire	x	ùxé	*tree
ɖ	ɔ́ɖɛ̀	*clan	h	ùhɛ̃́	knife
ɟ	ɔ́ɟánɖɛ́	gap in teeth	ʒ	ɔ́ʒá	poverty
g	ɔgɔ̀	hunger	j	ɔ́ja	portion of a farm
m	ámá	mother			

[1] p is rare.

SEKPELE (LIKPE) (54)

Phonological contrasts elicited from informant.

p	ɔpúnú	table	ɲ	ɔ́ɲə̀	you see
k͡p	k͡pɔk͡pɔ	duck	ŋ	àŋɔ̀nì	you write
t	ətə̀	you gave	f	ófò	he's got
t͡ɕ	ὲt͡ɕὲ	it matches	w	òwé	who
k	kɔnɔ̀	beautiful	v	ávò	you are free
b	əbə̀	you came	s	sìsi	yam
d	èdì	you eat	x	òxé	umbrella
g	ògàmɔ́	trap	h	òhìlò	*yam
m	ɛmɔ	it is big	j	éja	it has cracked
n	anɔ	you hear	l	ólà	message

SIWU (LOLOBI) (55)

Phonological contrasts elicited from the informant.

p	ìpã	scar	g	ígã́	toad
k͡p	ɔ̀k͡pɛ́	*herb	m	ìmã̀	blood
t	ìtá	stone	n	ɔ̀nã̀	he is absent
t͡ʃ	ɔ́t͡ʃὲ	he started	ɲ	àɲɛ́	breasts
k	àkéŋ	coal	f	ífɔ́	fear
kʷ	ɔ̀kʷàɛ́	soap	v	vaà	gaping wide
b	ɔ̀bí	child	s	ísã́	roof
g͡b	íg͡bã́	proverb	h	íhà	worry
d	adɔ̀	squirrel	j	ijó	house
ɖ	íɖà	fatigue	l	àlɛ́	you are good
d͡ʒ	ɔ́d͡ʒé	*name	ɹ	ìɹɔ́	something

KRACHI (56)

Phonological contrasts elicited from informant.

p	ɔ́pò	sea	n	àná	four
k͡p	ák͡pa	rubbish	nj	ànjɔ́	two
t	ɔ̀tὲ	he is living	ŋ	áŋὲ	let us cut
t͡ʃ	ɔ̀t͡ʃɛ́	female	ŋʷ	àŋʷà	*tree
t͡ʃʷ	ɔ̀t͡ʃʷɔ̀ɛ́	sun	ʍ	àʍéʍé	mirror
k	ɔ̀ké	tomorrow	f	ɔ̀fà	mother's brother
kʷ	ákʷɛ	tribal marks	w	ɔ̀wóɹé	book
b	ábɛ́	palm nuts	s	àsó	ears
d	ádù	medicine	ʃ	ɔʃɛ	force
d͡ʒ	ɔ́d͡ʒà	fetish priest	h	ahɪ́	annoying
d͡ʒʷ	íd͡ʒʷó	yams	j	ájú	millet
g	àgò[1]	silk	l	ɔlɔ	sore
m	ɔ̀mɔ̀né	nose	ɹ	ówùɹà	chief

[1] g is rare.

Gɛ̃ (57)

Phonological contrasts elicited from informant: nouns cited in definite form.

p	épè	year	m	èmɔ̃	instrument
k͡p	èk͡pé	stone	n	ènɔ̃	mother
t	étó	ear	ɲ	ɲónɔ̃	woman
t͡s	ètsɔ̀	mourning	f	fòfó	elder brother
k	èkɔ̀	neck	w	èwè	sun
b	ébɔ̀	he gathers	v	vɛ̀vɛ̀	*food
g͡b	ég͡bɔ̀	he comes back	s	ésó	he left
d	édó	he sows	x	éxá	broom
ɖ	eɖo	he has	h	èhà	song
d͡z	èd͡zè	salt	ɣ	éɣé	it is getting white
g	ègó	pocket	l	élɔ̀	he accepts

There are also a number of clusters with r or l alternating as in Ewe.

F5 (58)

Contrasts as indicated by Delafosse (1894).

k͡p	ák͡pà	wound	ɲ	g͡bèéɲó	happiness
t	àtá	bean cake	f	àfá	you are cold
t͡s	àtsɨ́ɔ́	you chose	w	àwà	arm
k	àkásá	cassava	v	àvɔ̀	cloth
b	ébú	he is lost	s	àsá	thigh
g͡b	ág͡bà	box	x	àxã̀	house
d	àdã̀	rashness	z	àzá	house
d͡z	àd͡zà	flank	zɥ	àzɥì [1]	fox
g	àgà	above	j	àjà	comb
m	àmà	remedy	l	àlè	income
n	ànà	you go . . .			

[1] zɥ is rare.

YORUBA (59)

Phonology in accordance with Siertsema (1958a & b, 1959). Compare also Abraham (1958), Olmsted (1953) and Ward (1952).

k͡p	ak͡pá	arm	ŋ	ŋólɔ	I shall go
t	ata	pepper	f	ɔfà	arrow
k	àká	barn	w	àwa	we
b	abà	village	s	asà	shield
g͡b	aàg͡bà	old	ʃ	àʃà	custom
d	àdá	cutlass	h	ahɔ́	tongue
ɟ	aɟá	dog	j	aja	wife
g	àga	chair	l	àlá	dream
m	amɔ̀	clay	ɾ	aɾa	body
n	àná	yesterday			

l and n may be in complementary distribution with reference to nasalized syllables (see page 23).

BINI (60)

Phonological analysis based on Melzian (1937), but with a different interpretation of nasalized w, j, etc. Compare also Wolff (1959).

p	opíà	cutlass		ʋ	eʋà	there
k͡p	òk͡pìá	man		v	evá	two
t	àtá	*tribe		s	ɛséi	generosity
k	ákò	*fruit		x	oxá	story
b	ɔ́bá	king		h	ohá	bush
g͡b	ɔg͡bá	pipe		z	ɛ̀zɛ̀	river
d	ádá	third		j	ɔjɔ́	rubber tap
g	ògì	melon		ɣ	oɣo	goat
m	èmà	drum		ɹ̥	àɹ̥à	*burial ceremony
n	oná	design		l	álázi	*monkey
f	ɛ́fɛ̀	wealth		ɹ̴	aɹ̴á	caterpillar
w	ewá	mat		ɹ	áɹába	rubber

ITSEKIRI (61)

Phonology as stated by the informant (A. Opubor) in personal communications. Compare Wolff (1959) for a similar analysis.

p	ípi	masquerade		n	aná	one bargains
k͡p	àk͡pà	fool		ɲ	áɲà	one sucks
t	ata	pepper		ŋ	áŋɔ̀	one snores
k	aká	kitchen shelf		f	afá	one scrapes
kʷ	ókʷa	he has gone off		w	awá	one comes
b	ɔbɛ	knife		s	asá	one dries
g͡b	ag͡bá	drum		ʃ	aʃɔ	cloth
d	ɔda	paint		j	ája	one writes
ɹ	áɹà	one fights		ɣ	áɣa	one scrapes
g	aga	chair		l	alá	one licks
gʷ	agʷá	one mixes		r	árà	one buys
m	áma	one moulds				

l and n are in complementary distribution with reference to nasalized syllables (see page 23).

URHOBO (62)

Phonological contrasts elicited in consultation with the informant. Compare Wolff (1959).

p	úpe	scar		ʍ	èʍá	bed
k͡p	òk͡pè	ugly person		f	éfã̀	others
t	òtá	talk		w	éwé	ponds
c	écá	coming		wᴸ	òwᴸó	soup
k	ɔ́kà	maize		ʋ	eʋá	fortune-telling
kʷ	ɔ́kʷa	let him pack		v	èvã̀	monitor lizards
b	ɔ́ba	king		s	ɔ̀sè	father
g͡b	ogba	fence		ʃ	èʃà	grey hair
d	òdà	cutlass		x	ɔ́xɔ̀	chicken
ɹ	oɹa	soap		z	òze	basin
g	ɔ̀gà	illness		ʒ	oʒa	suffering
gʷ	ɔ́gʷà	*yam		j	ɔ́já	journey
m	ɔ́mɔ̀	child		ɣ	ɔ̀ɣɔ	respect
ŋm	àŋmá	cloth		l	òlɔ̀	grinding-stone
n	ɔ̀nà	this		ɹ̴	ɔɹ̴e	tsetse fly
ɲ	eɲa	saliva		ɹ̥	ɔ́ɹ̥ɛ̀	plantain

ISOKO (63)

Phonological contrasts suggested by the informant who was a member of the local literacy committee. Compare also Hubbard (n.d. 1951?) and Wolff (1959) for slightly different analyses.

p	épe	spot	ʍ	òʍɛ́	laugh
k͡p	èk͡pè	halved calabash	f	òfɛ̀	razor
ṭ	òṭú	louse	w	ìwájò	trick
t	òtú	gang	wɪ	ɛ́wɪɛ́	hoe
k	ókò	creek	ʊ	ɛ́ʊɛ́	breath
b	ɛ̀bà	fish	v	ɛ́vɛ́	how?
g͡b	ɛ̀g͡bà	charm bracelet	s	úsì	line
ḍ	úḍù	farm	h	òhà	saying
d	údù	chest	z	úzì	proclamation
ɟ	óɟò	*tree	j	ìwájò	trick
g	ɔ̀gɔ̀	bottle	ɣ	òɣà	branch
m	òmà	body	l	òlá	jump
ŋ	óṇá[1]	walk	ɭ	òɭá	flight
n	òna	skill	ɾ	òɾá	yours
ŋʷ	òŋʷà	bone	ɽ̥	ɔ́ɽ̥a	schism

[1] ṇ = ɲ for some informants.

ORA (64)

Contrasts elicited from informant.

p	ɔ́píà	cutlass	f	ɛ̀fè	wealth
k͡p	ɔ̀k͡pà	cock	w	áwà	dog
t	àtà	truth	ʊ	ɛ́ʊóró	*town
tʃ	ótʃè	he calls	v	ɛ̀vɔ̃	*grass
k	àkò	sheath	s	ɛ́sɛ̀	greeting
b	óbɔ̀	arm	x	ɔ̀xà	story
g͡b	àg͡bɔ̀	people	h	ɔ̀hà	wife
d	àdà	witches' meeting-place	z	ózì	crab
			j	héjè	no
dʒ	òdʒè	chief	l	álálò	ringworm
g	ɔ̀gɔ̀	bottle	ɭ	ɛ́ɭáɭè	men in an extended family
m	òmò	soup			
n	ònà	this	ɾ	èɾà	father

IGBIRA (65)

Phonology in accordance with Ladefoged (1964).

p	àpápà	maize	ŋ	aŋɛ̀rɛ́	green
t	atá	pebbles	ŋʷ	aŋʷɛ́	palm oil
c	ɩcáca	grasshopper	ʍ	aáʍɛ́	hen
k	aká	charcoal	w	ɛwà	grey hair
b	baàba	thoroughly	v	avɔ	thanks
d	àdá	father	s	àsɛ̀	palm fruit
ɟ	aɟá	baby wrap	h	àhà	good fortune
g	àgá	chair	z	ɔ̀zà	people
m	àmá	but	j	ɔ́já	round worm
n	anɔ́	salt	r	àra	sleep
ɲ	aɲá	blood			

58

APPENDIX B

IGBO (67)

Phonological analysis based on Carnochan (1948) and Ward (1936). This form of Igbo (that of the villages near Owerri) has more contrasts than many other forms.

ph	àphé	sharpening	ɟ	áɟà	pulling apart
phj	áphjá	whipping	gɦ	àgɦá	being useful
p	ápè	pressing	g	àgá	walking
k͡p	àk͡pà	bag	gʷ	àgʷà	beans
th	àthá	blaming	m	ámá	village
t	àtá	chewing	n	ánú	meat
ch	áchà	cutting sharply	ɲ	áɲá	eye
c	áca	clearing a way	ŋ	áŋáà	cane
kh	ákhá	long	ŋʷ	áŋʷà	trying successfully
k	áká	hand	ʍ	áʍà	squeezing wet leaves
khʷ	ákhʷà	demolishing	f	áfɔ	year
kʷ	àkʷà	cloth	w	àwá	chop with axe
bɦ	mbɦá	boast	s	àsáà	seven
bɦj	ábɦjà	pressing heavily	ç	áçà	many
b	mbà	another town	h	áhà	boring needle
bj	ábjá	coming	z	ázà	sweeping
ɓɦ	àɓɦà	jaw	j	ájɔ	sifting
ɓ	àɓà	influence	ɣ	áɣá	war
dɦ	ádɦà	falling	l	àlà	ground
d	ádà	crab	ɹ	àɹá	eating fruit
ɟɦ	áɟɦà	earth	ɹj	áɹjà	making animal noises

IDOMA (WESTERN) (68)

Phonology in accordance with Armstrong (forthcoming).

p	àpà	lizard	n	ànà	*fruit
k͡p	àk͡pà	bridge	ɲ	áɲà	quick temper
t	áta	cricket	ŋ	ɔŋáɟi	Western rainbow
c	acá	wing	ŋʷ	àŋʷà	fortune-telling instrument
k	àka	wheel			
kʷ	ɔkʷɔ	*tree	f	àfà	blessing
b	àbà	palm nut	ɥ	ɔɥá	moon
g͡b	àg͡bà	jaw	w	ɔ́wá	redness
d	áda	father	s	òsì	answer
ɟ	àɟa	adultery	ʃ	oʃí	waist
g	àga	axe	h	àhà	walk
gʷ	àgʷa	swimming	j	ɔjá	width
m	áma	bell	l	álá	lamb
ŋ͡m	aŋ͡màa	painted body marks	r	ari	palm tree

IJAW (UPPER) (71)

Phonological contrasts elicited from the informant. Compare Wolff (1959) for a similar analysis.

p	àpà	consult them	ɲ	aɲaɲa	horse
k͡p	ak͡pa	bag	ŋ	àŋá	egg
t	àtá	go away	f	ɔfa	cord
k	àká	corn	w	àwɔ	children
b	àba	*fish	v	àvɔ	covenant
g͡b	àg͡bá	calabash	s	àsà	riverside
d	adà	coal tar	z	àzá	*fish
g	àgá	*yam	j	àjá	new
m	àmà	right	l	àlá	chief
n	anáà	haven't you?	ɺ	àɺá	your

KALABARI (72)

Phonological contrasts elicited from the informant. Compare Wolff (1959) for a slightly different analysis.

p	àpà	fool	g	ágá	cane
k͡p	àk͡pà	bag	m	àmà	right
t	áta	granny	n	ànà	*fish
k	áká	tooth	f	ɔfɔ	wand
b	àbà	*fish	w	àwɔ	children
ɓ	aɓa	*girl	v	válà[1]	sail
g͡b	àg͡bà	offering	s	ósó	contribute
d	àda	eldest daughter	j	ája	mother
ɗ	áɗa	father	l	álá	chieftaincy
ɺ	ójó	bronze	r	írí	stray

[1] v is rare.

KAMBARI (80)

Phonological contrasts suggested and elicited by C. Hoffmann. (^ denotes a falling tone.)

p	pàdâ	slaughter	n	à:ná:nà	log
k͡p	k͡pàtô	cover	f	fàbâ	beat
t	tàwâ	take away	w	wàlâ	go
ts	tsàâ	buy	ʔw	ù:ʔwâ	compound
tɕ	tɕá:mű	give me	v	vàzà	harnessed antelope
k	kàrâ	tear	s	sàrâ	throw
ʔ	sàʔâ	wash	ɕ	ɕàmâ	decay
b	á:bá:ba	indigo plant	z	zàzâ	scatter
ɓ	úɓàrâtò	he barked	ʑ	ʑàrâ	hit
g͡b	ìg͡bàg͡bâ	ducks	j	jàjâ	tickle
d	dàmmâ	say	ʔj	óʔjúwɔ́	thatching grass
ɗ	iɗàná[1]	lines	l	làpâ	beat, shoot
g	gàlâ	clear ground	r	ràwâ	throw
m	mà:nî	breast			

The one word ŋhwáʔá 'tomorrow' contains h and ŋ otherwise not accounted for here. There are also nasal compounds, and clusters with j and w.

[1] ɗ is sometimes ʔd.

EFIK (82)

Phonological contrasts in accordance with Ward (1933).

k͡p	àk͡pá	first	ŋ	sàŋá	walk
t	àtá	sixty	ŋʷ	áŋʷá	field
k	àkàm	prayer	f	àfàŋ	footpath
kʷ	ákʷá	wooden bowl	w	àwà	green
b	àbà	forty	s	àsábɔ	python
d	àdá	barren	j	áját	very hot
m	àmá	lover	ɣ	dàyù	go away
n	ànàŋ	eighty	r	bárá	set on fire
ɲ	áɲà	he sells			

TIV (83)

Phonological contrasts suggested and elicited by D. Arnott (cf. Arnott 1958).

p	á pèndà	he put on top	ndz	á ndzùùr	he muddled
k͡p	á k͡pèrà	he dragged	n̪d̪ʑ	á n̪d̪ʑɔ̀ɣɔ̀l	he spoke quickly
t	á tèsè	he showed	ŋg	á ŋgòhòr	he received
ts	á tsèɣà	he set apart	m	á mènà	he swallowed
tsʷ	á tsʷèr̀	he renounced	n	ánèndà	he is backward in
t̪ɕ	á t̪ɕɔ̀hɔ̀	he plastered			growth
k	á kèndè	he raised	ɲ	á ɲàndè	he urinated
b	á bèndè	he touched	f	á fèesè	he snatched up
g͡b	á g͡bèr̀	he slashed	w	á wèndè	he tied loosely
d	á dè	he left alone	v	á vèndà	he refused
dz	á dzèndà	he prohibited	s	á sèer̀	he increased
d̪ʑ	á d̪ʑìŋgè[1]	he searched	ʃ	á ʃàr̀	he sat down
g	á gèmà	he turned round	h	á hèmbà	it exceeded
mb	áa mbè	she suckled	j	á jèm	he went away
ŋ͡mg͡b	á ŋ͡mg͡bahom	he approached	ɣ	á tsèɣà	he set apart
nd	á ndèrà	he began	r	a ràm	he spoke

[1] d̪ʑ occurs in verbs before i only; in nouns before other vowels.
Palatalized consonants occur in a few verbs and in nouns with an i- prefix.

KUTEP (85)

Phonology based on Welmers (unpublished material). Not all possible clusters are noted here; in addition to a few more compounds with **w** (after occlusives) or **v** (after fricatives) and **ɕ** (after stops) or **k** (after fricatives) there are many palatalized consonants.

p	bapàp	they made	m	bamá	they went
pʷ	bapʷà	they grind	mʷ	bamʷà	they measured
k͡p	bak͡pák	they are strong	n	unà	the (?)stick
t	bata	they shot	ɲ	baɲàŋ	they are handsome
tʷ	batʷáp	they picked up	ŋ	baŋáŋ	they are red
ts	batsà	they prefer	ŋʷ	baŋʷáŋ	they slip
tsᶠ	batsᶠáp	they chose	f	bafa	they included
tç	ùrʷátçàk	his skin	fk	bafkap	they roasted
tʃ	batʃàk	they glean	w	bawam	they are dry
tɕᶠ	batɕᶠàk	they sleep	s	basà	they took
c	bacá	they found	sᶠ	basᶠa	they kneel
k	bakà	they share	sk	baskàp	they are better
kʷ	nsázᵛakkʷà	the water is hot	skʷ	baskʷáp	they are foolish
b	babá	they came	ʃ	baʃa	they found
bʷ	babʷa	they deceived	ɕᶠ	aɕᶠápaŋ	groundnuts
mb	bambà	they made a hole	z	bazap	[meaning obscure]
mbʷ	bambʷà	they tasted	zᵛ	nsázᵛakkʷa	the water is hot
nd	bandà	they gave	ʒ	baʒap	they bought
ndʷ	bandʷap	they wove	ʐᵛ	baʐᵛam	they begged
n̪d̪ʒ	mbàk͡pà ró	the maize	j	bajà	[meaning obscure]
	in̪d̪ʒa	germinated	r	bará	they are rotten
ŋg	baŋgak	they are quiet	rʷ	barʷa	they greeted
ŋgʷ	baŋgʷà	they drink			

NGWO (WIDEKUM) (87)

Phonological contrasts elicited from the informant.

p	épɛ́	cleanliness	n	énò	beads
k͡p	ekk͡pè	boat	ɲ	ɲem	meat
t	atø	head	ŋʷ	áŋʷèʔènè	book
ts	étsɔʔ	laughter	f	éfoʔ	cockroach
c	écé	medicine	w	ewy[1]	fire
k	ékúmú	name	s	àsame	twenty
ʔ	zeʔe	come	ʃ	eʃýný	gathering
b	ébùm	belly	z	èzɛ̰̀	vomit
g͡b	èg͡bɛ́	tattoo	ʒ	aʒ̰ì	shadow
d	édã́	loose weaving	j	àjé	woman
dz	èdzɛ̰̀	*dance	ɣ	ɣó	man
dʒ	dʒé	*fruit	ɬ	àɬá	hoe
ɟ	éɟɛ̰̀	postpone	l	àlà	thing
g	egɛ́	grasses	ɾ	ekèɾè	pepper
m	ámá	kernels			

There are also clusters with l preceded by labial and velar stops and fricatives; and with w preceded by labial, alveolar, post alveolar and velar stops, affricates and fricatives.
[1] Apparently in limited contexts only; may not be phonemic.

APPENDIX B

BAFUT (88)

Phonological contrasts suggested by the informant.

t	átì	tree	ɲ	ɲà[1]	insect
tʃ	atʃáʔə	arrogance	ŋ	nɔŋə́[1]	lie
k	ákɵ̀	forest	f	áfù	medicine
ʔ	awàʔà	*man	w	awàʔà	*man
b	ábo	hand	s	sá	drinking season
d	ádì	malicious act	z	ázi	he is coming
dʒ	ád̠ʒɵ̀	plum	j	ajà	path
m	mó	child	ɣ	áɣɯ	receptacle
n	nú	honey	l	alò	year

There are also several clusters with w as the second element.

[1] ɲ and ŋ seem to be in complementary distribution.

KOM (89)

Some material has been presented by Bruens (1947); but it is not sufficient for a phonological analysis. My own data are also very limited. There are certainly many phonemes and combinations of phonemes not illustrated in the following list of contrasts, some of which may, in any case, be allophonic.

t	átú	head	gᵛ	sɯ̀gᵛà	to grind
tɕ	àtɕɯ́	seat	gɣ	ŋ̀gɣò	antelope
tɕ'	ɯtɕ'u	mouth	m	kamàkɔ́	I want to go up
k	sɯ̀kɔ̀ŋ	to love	n	ńdó	house
k'	ìk'ú	death	ɲ	ɯ̀ɲám	animal
ʔ	ɯŋʷaʔlɯ	book	ŋʷ	ɯ̀ŋʷèn	him
b	íbál	level land	f	ìfə̀f	blindness
bz	ibzi	giving birth	s	ísáŋ	maize
bɣ	íbɣál	shouting	ʒ	sɯ̀ʒɯ̀	to eat
d	ídíàŋ	crossing	ɣ	aɣó	matters
dʑ	údʑùŋ	good condition	jᵛ	ajᵛus	spirit
dɣ	sɯ̀dɣàl	to swank	l	áfɔ́ŋlɯ́	valley
g	ŋ̀goʔ	white ant			

63

HAUSA (91)

Phonological contrasts suggested and elicited by D. Arnott. Compare Greenberg (1941), Hodge (1947), James and Bargery (1925) and Carnochan (1952) for different analyses.

t	táːʃiː	departure	gʷ	gʷándàː	pawpaw
tʃ	tʃíkì	stomach	m	màgánàː	palaver
c	câŋʷáː	cat	n	náːmàː	meat
c'	c'áːlèːwáː	oversight	f	fàːɽáː	locust
k	kàːsúwáː	market	fj	fjàːʔdáː	flogging
k'	k'óːfàː	doorway	w	wútáː	fire
kʷ	kʷálábáː	bottle	s	sàndáː	stick
kʷ'	kʷ'àːɽóː	insect	s'	s'ábtàː	cleanliness
ʔ	ʔúwáɽsà	his mother	ʃ	ʃáːnú	cattle
b	báːɽíkì	barracks	h	háwáː	riding
ʔb	ʔbàɽáːwòː	thief	z	zúmàː	bees
d	dáːwà	guinea corn	j	jáːɽòː	boy
ʔd	ʔdínkì	sewing	ʔj	ʔjáːʔjáː	children
dʒ	dʒáː	red	l	lóːkàt̪ʃíː	time
ɟ	ɟàʔdáː	groundnuts	ɽ	báɽàː	servant
g	gàɽíː	town	ɾ	báɾàː	begging

BURA (92)

Phonology in accordance with Hoffmann (personal communication; cf. Hoffmann, 1957 [not seen]; cf. also Wolff (1959) for a different analysis).

p	pàkà	search	mj	mjà	talk
ps	psá	lay eggs	mʷ	mʷàntà	remove
pʃ	pʃàrì	spread a net	m͡np͡t	m͡np͡tà	die
pʷ	pʷá	alight	ɲ	ɲárá	lick
p͡t	p͡tá	*animal	ŋ	ŋáŋá	bird
p͡ts	p͡tsà	roast	ʍ	ʍàdá	groundnuts
p͡tʃ	p͡tʃi	sun	f	fà	take
t	tá	cook	fʷ	fʷàrì	forget
ts	tsá	weave	w	wáskí	each
tʃ	tʃàrà	choose	ʔw	ʔwálá	big
c	càŋɟàr	tongue	wᴸ	wᴸàrwᴸar	throat
k	kálá	bite	v	ávánà	this year
kʷ	kʷárá	donkey	s	sà	drink
b	bàrà	want	ʃ	ʃá	lost
bz	bzáŋa	brother	ç	çál	intestine
bʷ	bʷà	part	h	hàlà	get old
ʔb	ʔbáɬà	dance	z	zàmtà	spoil
ʔbʷ	ʔbʷá	beat	ʒ	ʒàrì	desist
ʔbd̪	ʔbd̪à	chew	j	jà	give birth
d	dàwà	enemy	ʔj	ʔjáhà	doctor
ʔd	ʔdà	eat meat	ɣ	ɣà	backbite
dz	dzàmà	think	ɬ	ɬá	cow
dʒ	dʒàrá	palm-oil tree	ʎ	ʎálá	cucumber
ɟ	ɟàl	bull	l	là	build
g	gàm	ram	ɮ	ɮábʷá	beat
gʷ	gʷàr	poison	ʌ	wúʌá	neck
m	màrà	carve	r	rútá	think little of

64

MARGI (93)

Phonological contrasts taken from a MS copy of Hoffmann (1963); elicited with help from A. Opubor speaking Hausa. Hoffmann has a slightly different phonetic specification.

p	páʔdó	rain	d	dàlmà	big axe
ps	psár	grass	ʔd	ʔdàʔdàhꭃ	bitter
pɬ	pɬàt	make wider	dz	dzànì	know
pɕ	pɕír	tree	d͡ʒ	d͡ʒágʷa	fish
pʷ	pʷa	pour in	nd	ndàl	throw
p͡t	p͡təl	chief	ndz	ndzə̀ndzə̀ʔbó	covered
p͡ts	p͡tsàp͡tsà	roasted	nd͡ʒ	nd͡ʒà	open wide
p͡tʃ	p͡tʃíp͡tò	washed	ɟ	ɟáʔdí	hump (of cow)
mp	mpà	fight	ɲɟ	ɲɟárí	leave
m͡np͡t	m͡np͡tàgò	bush	g	gàlí	spear
m͡np͡ts	m͡np͡tsàkù	pick up	gʷ	gʷà	enter
m͡np͡tʃ	m͡np͡tʃàʔdò	flute	ŋg	ŋgàdó	remain over
t	tátá	that one	ŋgʷ	ŋgʷal	be bent
ts	tsá	beat	vᵇ	bóvᵇú	[ideophone]
tʃ	tʃátʃá	all	m	màlà	woman
tʷ	tʷàŋ	log of firewood	mʷ	mʷàlmʷàl	sour
nt	ntà	split	m͡n	m͡nà	mouth
nts	ntsàntsà	shouted	n	náná	locust bean tree
ntʃ	ntʃà	point at	ɲ	ɲàɲafꭃ	sprinkled
ntʷ	ntʷá	split in two	ŋ	ŋàŋà	bent
ntsʷ	ntsʷa	thread	ŋʷ	ŋʷà	face
c	cácíkà	whooping cough	ʌ	ʌàʌà	boiled
ɲc	ɲcàhò	break (calabash)	f	fà	farm
k	kákádò	book	fʷ	fʷà	pool
kʷ	kʷàkʷá	weak person	w	káwà	sorry
ŋk	ŋkà	refuse	ʔw	ʔwáʔwí	adornment
ŋkʷ	ŋkʷà	girl	wᶦ	wᶦá	reach inside
ʔ	áʔà	no	v	vál	granary
b	bábál	open place	vʷ	vʷànd͡ʒà	*frog
bz	bzɘ́r	child	s	sà	drink
bʒ	bʒɘ̀nà	kill many	sʷ	sʷá	shut
bj	bjábjágó	small bat	ʃ	ʃàʃílgà	star
bʷ	bʷál	ball	ç	çà	moon
b͡d	b͡dàgò	valley	x	xá	big water pot
b͡dz	b͡dzà	foolish	z	zárzár	always
b͡d͡ʒ	b͡d͡ʒàgò	*compound	ʒ	ʒábá	re-sow
mb	mbà	tie	j	jà	give birth
mbʷ	mbʷàʃkà	forearm	ʔj	ʔjà	thigh
m͡nb͡d	m͡nb͡dá	surpass	jᶦ	jᶦàjᶦàʔdò	picked up
m͡nb͡dz	m͡nb͡dzànì	spoil	ɣ	ɣàfó	arrow
m͡nb͡d͡ʒ	m͡nb͡d͡ʒahꭃ	flute	ɬ	ɬànà	cut off
ʔb	ʔbàʔbàl	hard	ɬʷ	ɬʷa	cut in two
ʔbʷ	ʔbʷàʔbʷà	cooked	ɮ	ɮà	fall
ʔb͡d	ʔb͡dàʔb͡dò	chewed	l	là	dig
			r	gʷàrà	play

SONGAY (95)

Examples from Prost (1956).

t	ta	receive	ɲ	ɲa	mother
tₔ	tₔa	call to prayer	ŋ	ŋa	eat
k	ka	[modal auxiliary verb]	f	ma:fa	sauce
			w	wa:	milk
b	ba	desire	s	sahã	power
d	da:	luck	h	hãsi	dog
dʐ	dʐadʐe	spend the evening with	z	za	take in the hand
			j	ja	there
g	ga:	body	l	bolo	echo
m	ma	understand	r	bo:ro	redden
n	nama	murder			

KRIO (96)

Phonology according to E. Jones (personal communication).

p	pan	on top	ŋ	ɛŋ	hang
k͡p	k͡pák͡pákùʁù	dried bonga	f	fɛt	fight
t	tap	stop	w	wahálà	worry
tʃ	tʃap	chop	v	vɛks	vex
k	kàŋgà	magic	s	sɛf	self
b	ban	band	ʃ	ʃák͡pá	sorrel
g͡b	g͡bàtɛ̀	concern	z	kɔ̀zín	cousin
d	du	do	ʒ	plɛ́ʒɔ̀	pleasure
dʒ	dʒab	job	j	jàg͡bá	fuss
g	go	go	l	lɛk	like
m	mas	cat	ʁ	ʁak	fight
n	nak	knock			
ɲ	ɲáɲám	food			

BIBLIOGRAPHY

Abercrombie, David (1953) Phonetic transcriptions, *Le Maître Phonétique*, 3d. Series, 100, 32–4.

Abercrombie, David (forthcoming) *Elements of general phonetics.*

Abraham, R. C. (1935) *The principles of Idoma*, London: Crown Agents, Lagos: C.M.S.

Abraham, R. C. (1940) *Principles of Tiv*, London: Crown Agents.

Abraham, R. C. (1958) *Dictionary of Modern Yoruba*, London: University of London Press.

Ansre, G. (1961) *The Tonal Structure of Ewe*, M.A. Thesis presented 1961, The Hartford Seminary Foundation.

Arnott, D. W. (1958) The classification of verbs in Tiv, *Bull. School of Oriental and African Studies*, 21.1, 111–133.

Bazin, Hippolyte (1906) *Dictionnaire bambara-français*, Paris: Impr. Nationale.

Bell, A. M. (1867) *Visible Speech*, London.

Berry, J. (1951a) *Pronunciation of Gã*, Cambridge: Heffer.

Berry, J. (1951b) *The Pronunciation of Ewe*, Cambridge: Heffer.

Berry, J. (1955) Some notes on the phonology of the Nzema and Ahanta dialects, *Bull. School Oriental and Afr. Stud.*, 17.1, 160–5.

Berry, J. (1957) Vowel Harmony in Twi, *Bull. School of Oriental and African Studies*, 19, 124–130.

Berry, J. (1959) The structure of the noun in Kisi, *Sierra Leone Studies*, 12, 308–15.

Bloch, B. and G. Trager (1942) *Outline of Linguistic Analysis*, Linguistic Society of America, Special Volume.

Bruens, A. (1947) The structure of Nkom and its relations to Bantu and Sudanic, *Anthropos*, 37/40, 4/6, 1942–45.

Carnochan, J. (1948) A study in the phonology of an Igbo speaker, *Bull. School of Oriental and African Studies*, 12.2, 417–26.

Carnochan, J. (1952) Glottalization in Hausa, *Trans. Phil. Soc.*, 79.

Carnochan, J. (1960) Vowel harmony in Igbo, *African Language Studies*, 1, 155–63.

Catford, J. C. (1939) On the classification of stop consonants, *Le Maître Phonétique*, 3d. Series, 65.

Catford, J. C. (forthcoming) *Phonation types*, (personal communication).

Chomsky, Noam. (1962) The logical basis of linguistic theory, *Preprints of papers for the IXth International Congress of Linguists*, Cambridge, Mass.

Christaller, J. G. (1881 [2nd ed. 1933]) *Dictionary of the Asante and Fante called Tshi* (*Twi*), Basel: Evangelishe Missionsgesellschaft.

Crosby, K. H. & Ida C. Ward (1944) *An Introduction to the Study of Mende*, Cambridge: Heffer.

Delafosse, Maurice (1894) *Manuel dahoméen*, Paris: Leroux.

Dunstan, M. E. (1964) Towards a phonology of Ngwe, *Jour. W. Afr. Langs.*, 1.1, 39–42.

Fant, Gunnar (1958) Modern instruments and methods for acoustic studies of speech, *Proceedings of the VIII Internat. Congress of Linguists*, Oslo: Oslo University Press.

Fant, Gunnar (1960) *Acoustic theory of speech production*, 'S-Gravenhage: Mouton.

Fischer-Jorgensen, E. (1958) What can the new techniques of acoustic phonetics contribute to linguistics? *Proc. of the VIII Internat. Congress of Linguists*, Oslo: Oslo University Press.

Greenberg, Joseph H. (1941) Some problems in Hausa phonology, *Language*, 17, 316–23.

Greenberg, Joseph H. (1963) The languages of Africa, *Internat. J. of American Linguistics*, 29.1, Part II.

Greenberg, Joseph (Ed. 1963) *Universals of Language*, Cambridge, Mass: M.I.T. Press.

Greffier, H. (1901) *Dictionnaire Français-Sérèr*, Ngasobil: Mission St. Joseph.

Hockett, Charles F. (1955) *A Manual of Phonology*, Baltimore: Waverly Press.

Hodge, Carleton T. (1947) An outline of Hausa grammar, *Language*, 23.4, Supplement.

Hoffmann, Carl (1957) Ph.D. Thesis, Hamburg.

Hoffmann, C. F. (1963) *A grammar of the Margi language*, Oxford: Oxford University Press.

Hubbard, J. (n.d. 1951?) *The Sobo of the Niger Delta*, Zaria: Gaskiya.

Hudgins, C. V. & R. H. Stetson (1935) Voicing of consonants by depression of the larynx, *Arch. Néerland. Phon. Expér.*, 11, 1–28.

IPA Principles (1949) *The Principles of the International Phonetic Association:* Department of Phonetics, University College, London.

Jakobson, R. & M. Halle (1956) *Fundamentals of Language*, 'S-Gravenhage: Mouton.

James, A. Lloyd and George P. Bargery (1925) A note on the pronunciation of Hausa, *Bull. School of Oriental and African Studies*, 3.4, 721–8.

Joos, M. (1948) Acoustic phonetics, *Language*, 24.2, Supplement.

Labouret, Henri (1952) *La langue des Peuls ou Foulbé*, Dakar: IFAN (Mém. 16).

Ladefoged, Peter (1957) Use of palatography, *Journal Speech & Hearing Disorders*, 22.5, 764–74.

Ladefoged, Peter (1960) The value of phonetic statements, *Language*, 36.3, 387–96.

Ladefoged, Peter (1962a) *Elements of Acoustic Phonetics*, Edinburgh: Oliver and Boyd.

Ladefoged, Peter (1962b) *The Nature of Vowel Quality*, Coimbra: Laboratorio da Phon. Exp.

Ladefoged, Peter (1962c) Sub-glottal activity during speech, *Proceedings of the IV Internat. Congress of Phonetic Sciences*, 'S-Gravenhage: Mouton.

Ladefoged, Peter (1963) in *Proceedings of the International Conference on Research Potentials in Voice Physiology* (Brewer, David .W, ed).

Ladefoged, Peter (1964) Igbirra notes, *Jour. W. Afr. Langs.*, 1.1, 27–37.

Lehiste, Ilse and Gordon E. Peterson (1961) Some basic considerations in the analysis of intonation, *J. Acoust. Soc. Amer.*, 33, 419–25.

Lehiste, I. (1962) Study of [h] and whispered speech, *J. Acoust. Soc. Amer.*, 34.5, 742.

Melzian, Hans J. (1937) *A concise dictionary of the Bini Language of southern Nigeria*, London: Kegan Paul.

Moll, Kenneth L. (1960) Cinefluorographic techniques in speech research, *J. Speech and Hearing Research*, 3, 227–41.

Olmsted, D. L. (1953) Comparative notes on Yoruba & Lucumi, *Language*, 29, 157–64.

Peterson, Gordon E. (1957) in *A Manual of Phonetics* (Kaiser, L., ed.), Amsterdam: North Holland Publishing.

Pike, K. L. (1943) *Phonetics*, Ann Arbor: The University of Michigan Press.

Pike, K. L. (1947) *Phonemics*, Ann Arbor: The University of Michigan Press.

Prost, André (1956) *La langue songhay et ses dialectes*, Dakar: IFAN (Mém. 47).

Risberg, Arne (1962) Fundamental frequency tracking, *Proceedings of the IV Internat. Congress of Phonetic Sciences*, 'S-Gravenhage: Mouton.

Rycroft, David (1960) Melodic features in Zulu eulogistic recitation, *African Language Studies*, 1, 60–78.

Senghor, Lèopold S. (1944) L'harmonie vocalique en Sérère (dialecte du Dyéguèome), *J. Soc. Africanistes*, 14, 17–23.

Siertsema, B. (1958a) Some notes on Yoruba phonetics & spelling, *Bulletin de l'Institut Français d'Afrique Noire*, XX, Series B, 576–95.

Siertsema, B. (1958b) Problems of phonemic interpretation 1. Nasalised sounds in Yoruba, *Lingua*, 7.4, 256–66.

Siertsema, B. (1959) Problems of phonemic interpretation 2. Long vowels in a tone language, *Lingua*, 8.1, 42–64.

Spencer, John (Ed. 1963) *Language in Africa*, Cambridge: Cambridge University Press.

Stevens, K. and A. S. House (1955) Development of a quantitative description of vowel articulation, *J. Acoust. Soc. Amer.*, 27, 484–93.

Sumner, A. T. (1921) *Handbook of the Sherbro language*, London: Crown Agents for Sierra Leone Govt.

Sumner, A. T. (1922) *A handbook of the Temne language*, Freetown: Govt. Printer.

Sweet, Henry (1906) *A primer of Phonetics*, Oxford: The Clarendon Press.

Sweet, Henry (1911) Phonetics, in *Encyclopedia Brittanica*, XI edition, 21, 458–67.

Truby, H. M. (1959) Acoustico-cineradiographic analysis considerations with special reference to certain consonantal complexes, *Acta Radiologica*, Suppl. 182.

Uldall, Elizabeth T. (1958) A high speed film of the human vocal cords, *Proc. VIII Internat. Congress of Linguists*, Oslo: Oslo University Press.

Voegelin, C. F. (1956) Linear phonemes and additive components, *Word*, 12, 429–33.

Voegelin, C. F. and F. M. Voegelin (1959) Guide for transcribing unwritten languages in fieldwork, *Anthropological Linguistics*, 1.6, 1–28.

Ward, Ida C. (1933) *The phonetic and tonal structure of Efik*, Cambridge: Heffer.

Ward, Ida C. (1936) *An introduction to the Ibo language*, Cambridge: Heffer.

Ward, Ida C. (1939a) A short phonetic study of Wolof (Jolof) as spoken in the Gambia and in Senegal, *Africa*, 12.3, 320–24.

Ward, Ida C. (1939*b*) *The Pronunciation of Twi*, Cambridge: Heffer.

Ward, Ida C. (1952) *An introduction to the Yoruba Language*, Cambridge: Heffer.

Weiss, H. (1939) Grammaire et lexique diola du Fogny, *Bulletin de l'Institut Français d'Afrique Noire*, 1.2–3.

Welmers, William E. (1946) A descriptive grammar of Fanti, *Language*, 22.3, Supplement.

Welmers, William E. (1950) Notes on two languages in the Senufo group, *Language*, 26.1, 126–46.

Welmers, William E. (1952) Notes on the Structure of Bariba, *Language*, 28.1, 82–103.

Westermann, Diedrich H. (1930) *A Study of the Ewe Language*, [trans. by A. L. Bickford Smith] Oxford: Oxford University Press.

Westermann, D. & M. A. Bryan (1952) *Languages of West Africa*, Oxford: Oxford University Press.

Westermann, D. & Ida C. Ward (1933) *Practical phonetics for students of African Languages*, Oxford: Oxford University Press.

Wilson, W. A. A. (1961) *An outline of the Temne language.* London: School of Oriental and African Studies.

Wolff, Hans (1959) Subsystem typologies and area linguistics, *Anthropological Linguistics*, 1.7, 1–88.

INDEX